THE LANGUAGE OF CELLS

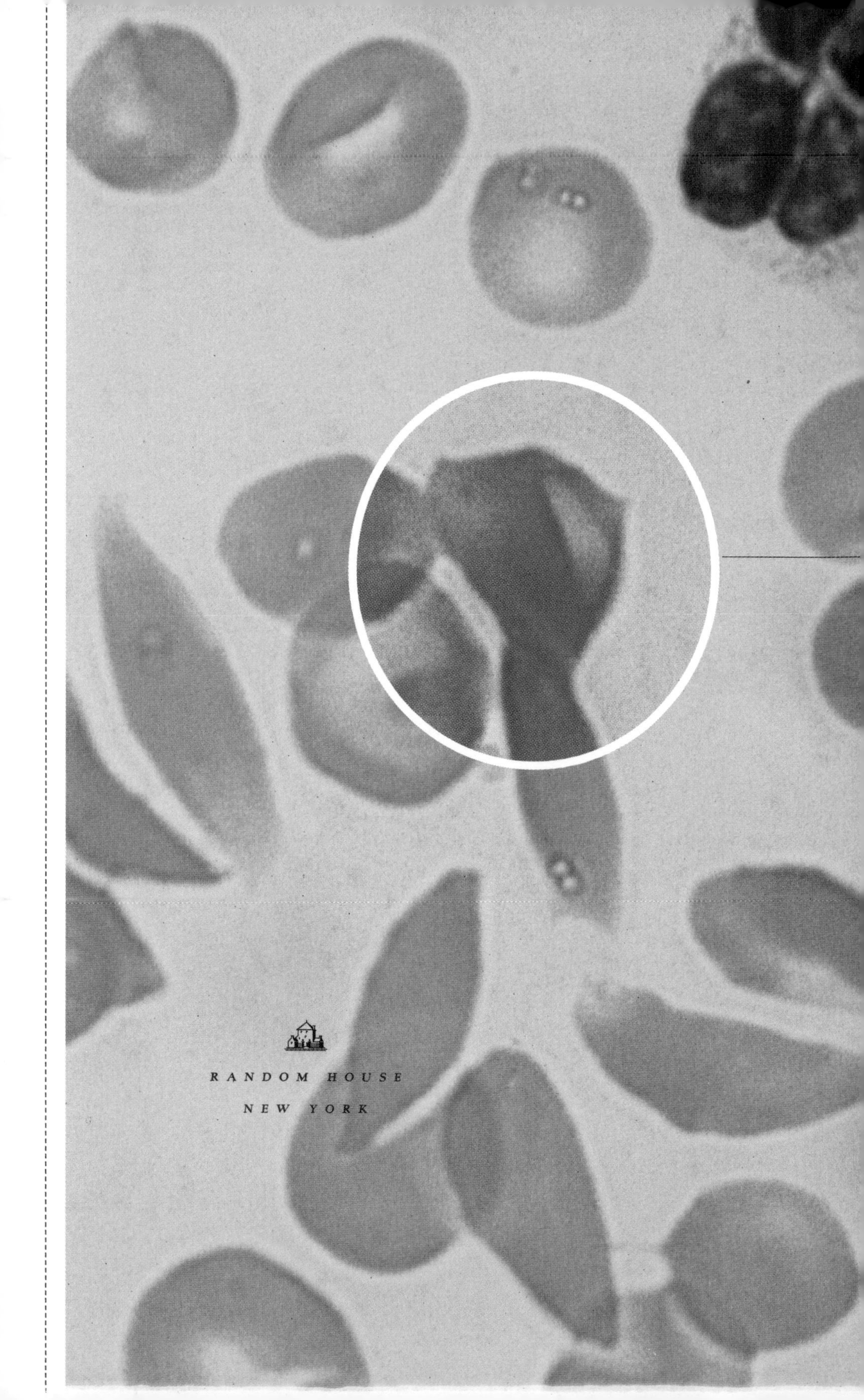

RANDOM HOUSE
NEW YORK

THE *Language of Cells*

LIFE AS SEEN UNDER THE MICROSCOPE

Spencer Nadler

Some of the essays in this work have been previously published in *The American Scholar, Harper's, The Massachusetts Review, The Missouri Review,* and *The Pacific Review*. In addition, "Brain Cell Memories" appeared in *The Best American Essays 2001* and "An Old Soldier" appeared in *The Best American Essays, 1999*.

LIBRARY OF CONGRESS CATALOGING-IN-PUBLICATION DATA
Nadler, Spencer.
The language of cells: life as seen under the microscope/Spencer Nadler.
p. cm.
ISBN 0-375-50416-8 (hardcover : alk. paper)
1. Pathology, Surgical—Anecdotes. I. Title.
RD57 .N33 2001
616.07—dc21
00-067369

Printed in the United States of America on acid-free paper
Random House website address: www.atrandom.com

9 8 7 6 5 4 3 2

First Edition

Book design by Barbara M. Bachman

For Myra

In loving memory of
Hyman "Kelly" Nadler

Whole lives are spent in the tremendous
affairs of daily events without even approaching
the great sights that I see every day.

WILLIAM CARLOS WILLIAMS
"THE PRACTICE"

ACKNOWLEDGMENTS

Foremost, I wish to thank those who willingly and patiently shared the stories of their illnesses with me in the hope of helping and inspiring others. I owe them a special gratitude that is more meaningfully expressed in the chapters of this book.

I particularly thank my dear friends and physician colleagues Jerry Finklestein and Richard Ellis for lively and constructive discussions and their unwavering support and encouragement, and Roger Terry, Emeritus Professor of Pathology at USC and surgical pathologist extraordinaire, for his careful scrutiny of the manuscript. Any errors of omission or commission are, of course, entirely my own.

I am indebted, too, to the readers of early chapters: Toni Frank, Amy Gerstler, Hideo Itabashi, Hope Edelman, Danny Miller, Bernd Scheithauer, Dan Silverman, and Deborah Lott; and final book versions: Colin Harrison, Bill Clark, Lynne Shook, Karen Kasaba, Virginia Holmquist, Teddi Softley, Paula Groncy, Phyllis Page, Jill Shapira, and David Lefkowitz.

To my editor, Kate Medina, and her assistant, Frankie Jones, a heartfelt thanks for their devotion to the editorial task and their belief in the book.

And, finally, kudos to my agent, Kris Dahl, whose abiding efficiency from the start has kept this project unencumbered.

CONTENTS

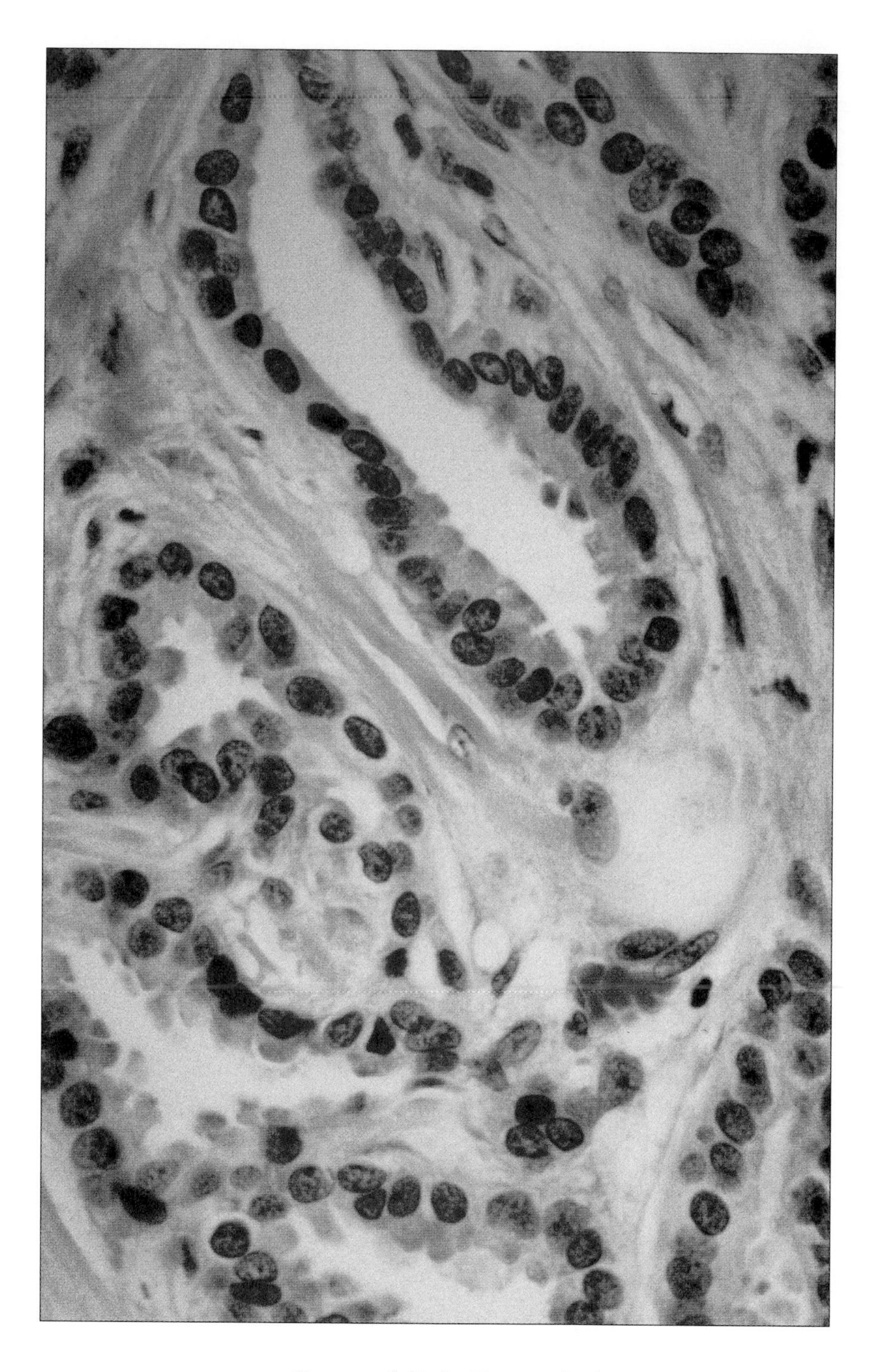

CHAPTER 1: *Distorted Hula Hoops of a breast cancer.*

COURTESY OF RICHARD ELLIS, M.D.

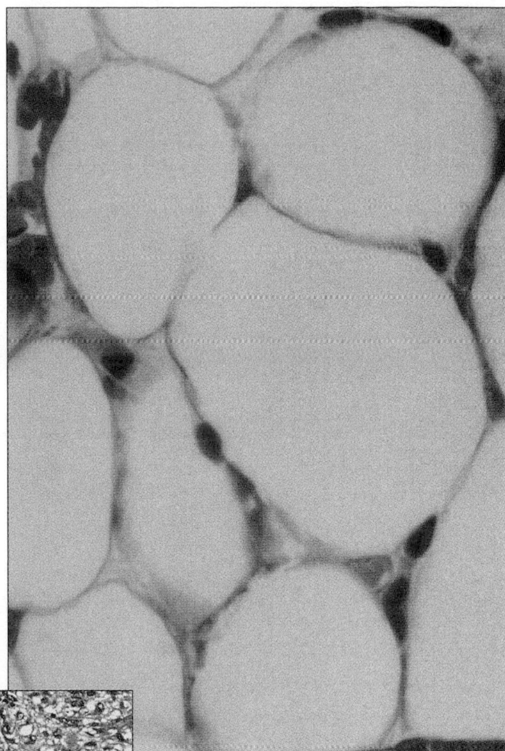

CHAPTER 2:

Blue nuclei and pink cytoplasm of fat cells pushed to a rim.

COURTESY OF RICHARD ELLIS, M.D.

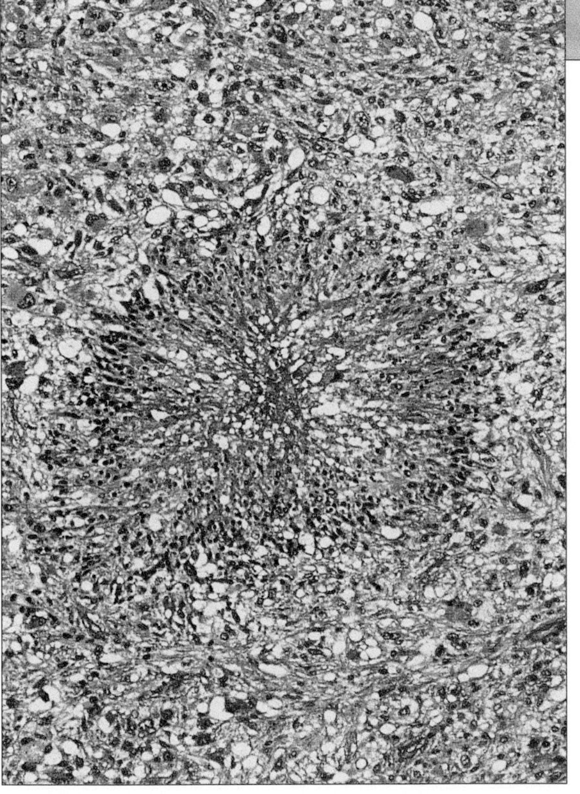

CHAPTER 3:

Brain tumor (glioblastoma): A palisade of tumor cells surrounds dead brain.

COURTESY OF JOEL CHEN, M.D.

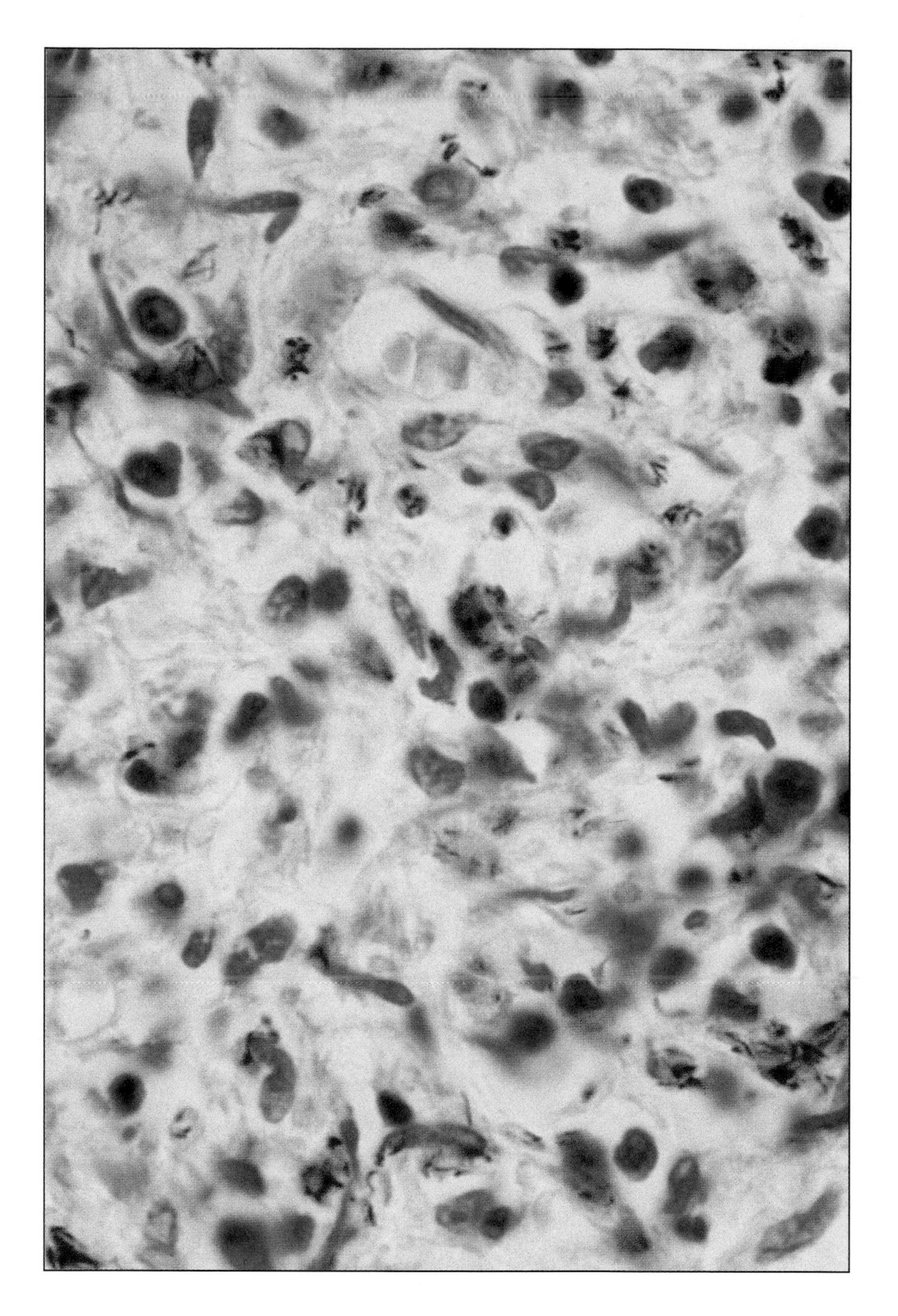

CHAPTER 3: *Blood-red tuberculosis bacilli.*

COURTESY OF THE McGRAW-HILL COMPANIES

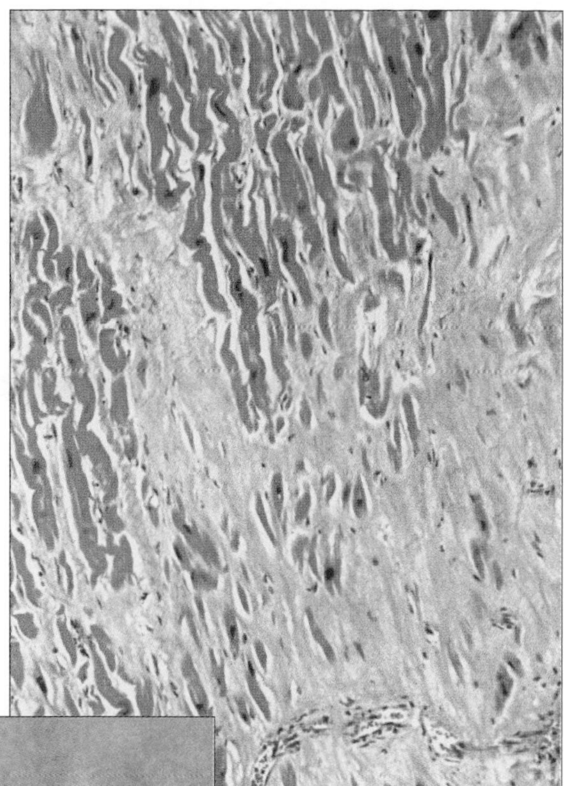

CHAPTER 4:
Scar tissue replacing ruddy myocytes of heart.

COURTESY OF THE McGRAW-HILL COMPANIES

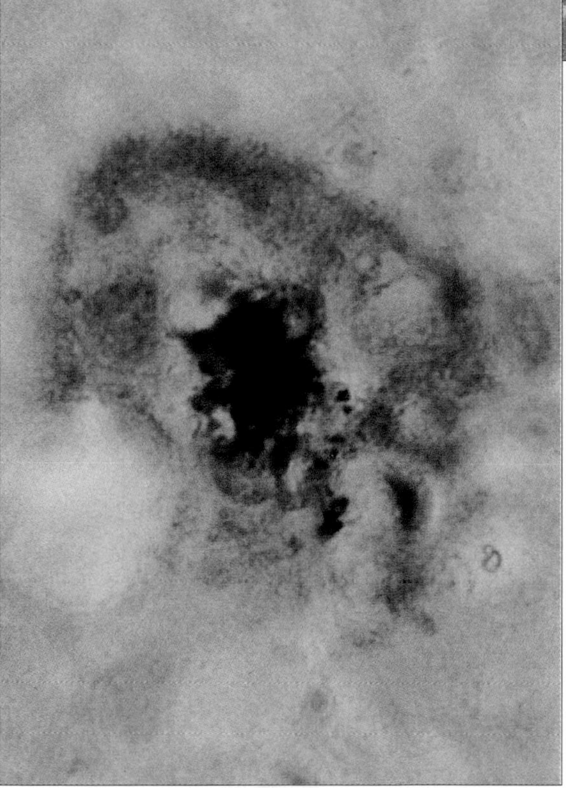

CHAPTER 5:
Bull's-eye of a classic Alzheimer's plaque.

COURTESY OF HARRY V. VINTER, M.D.

CHAPTER 6: *Blood smear of sickle cells.*

COURTESY OF RICHARD ELLIS, M.D.

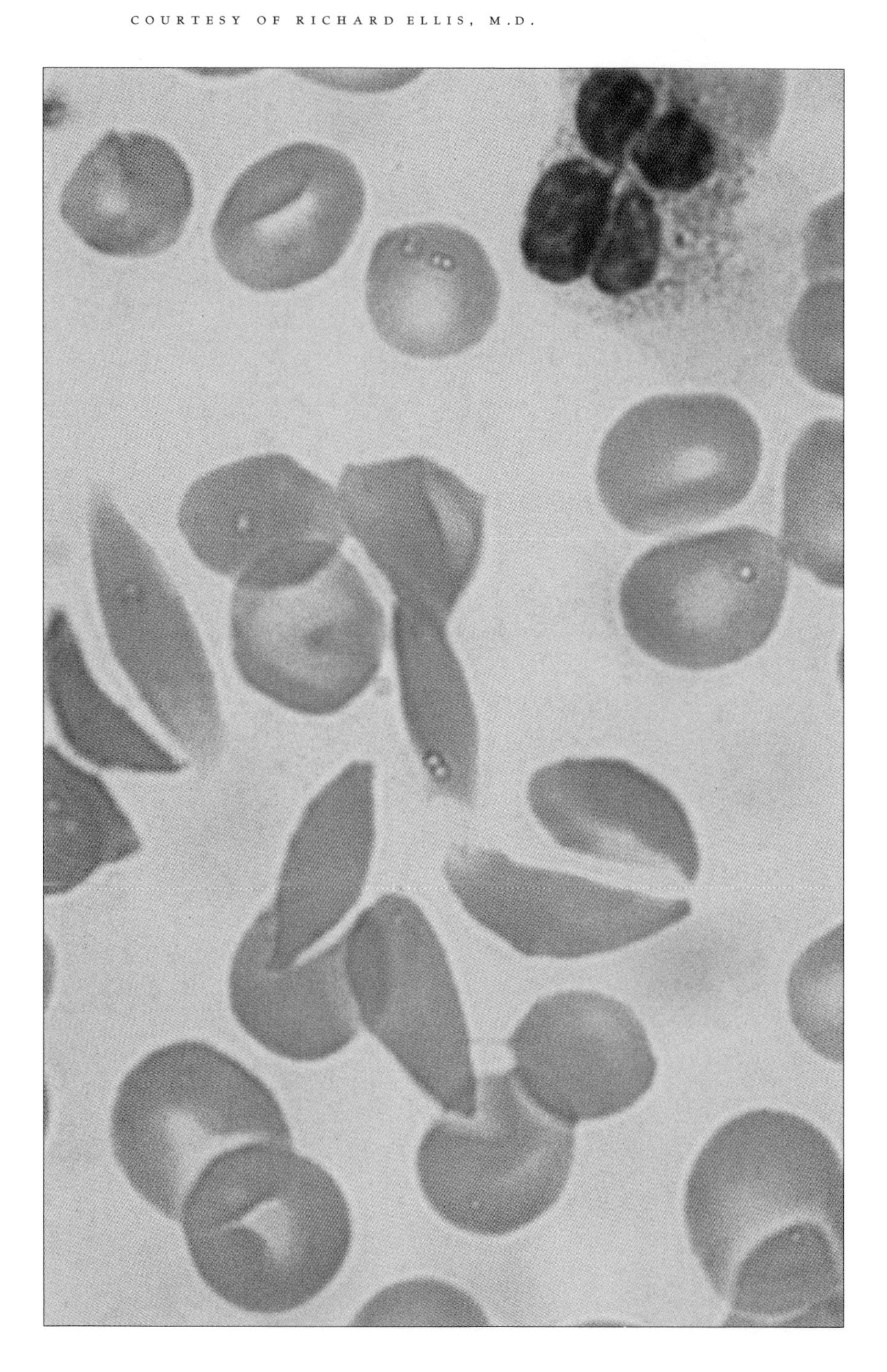

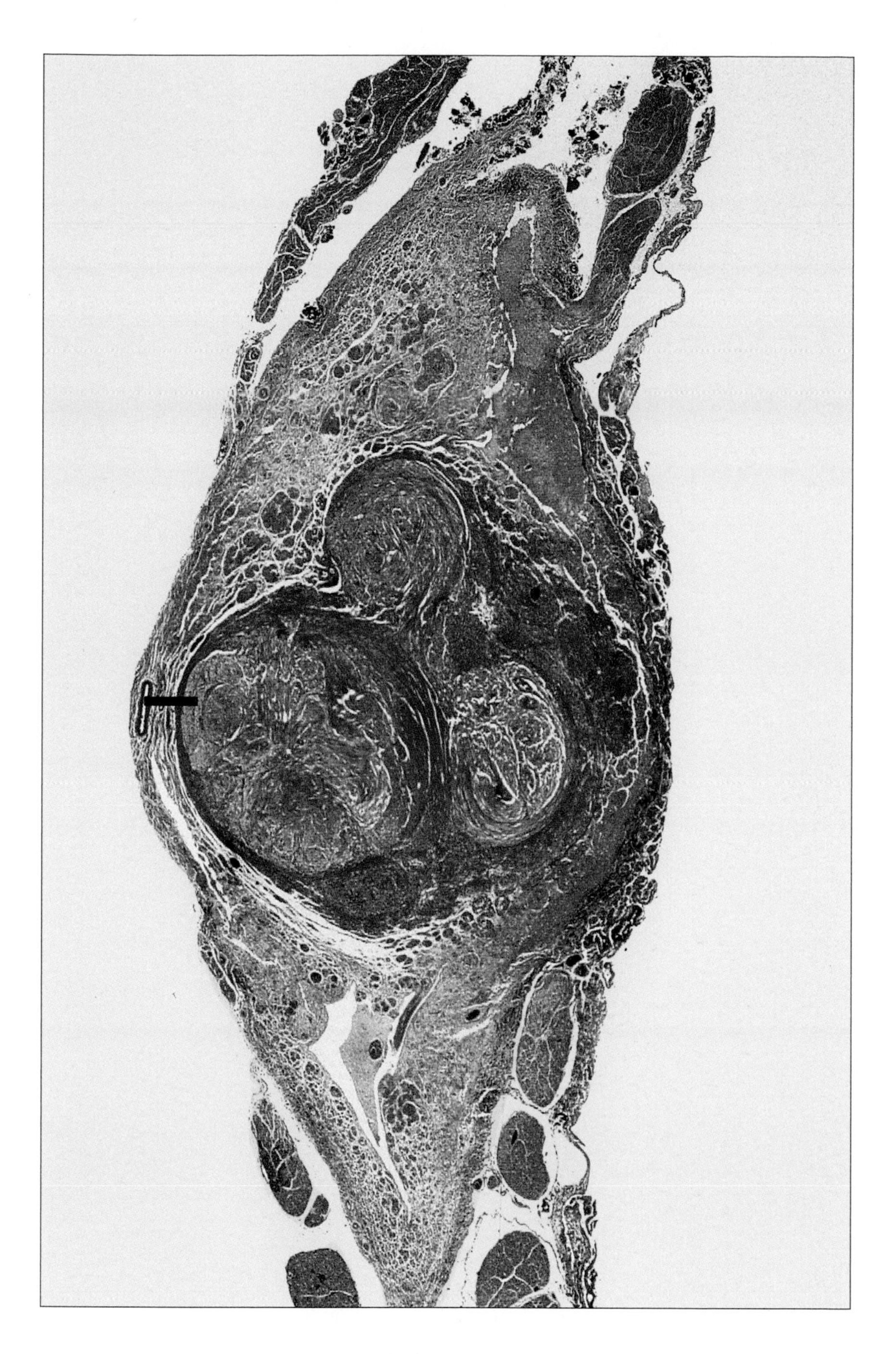

CHAPTER 7: *Severed edge of a paraplegic spinal cord showing benign tumor (neuroma) bound in scar.*

AUTHOR'S COLLECTION

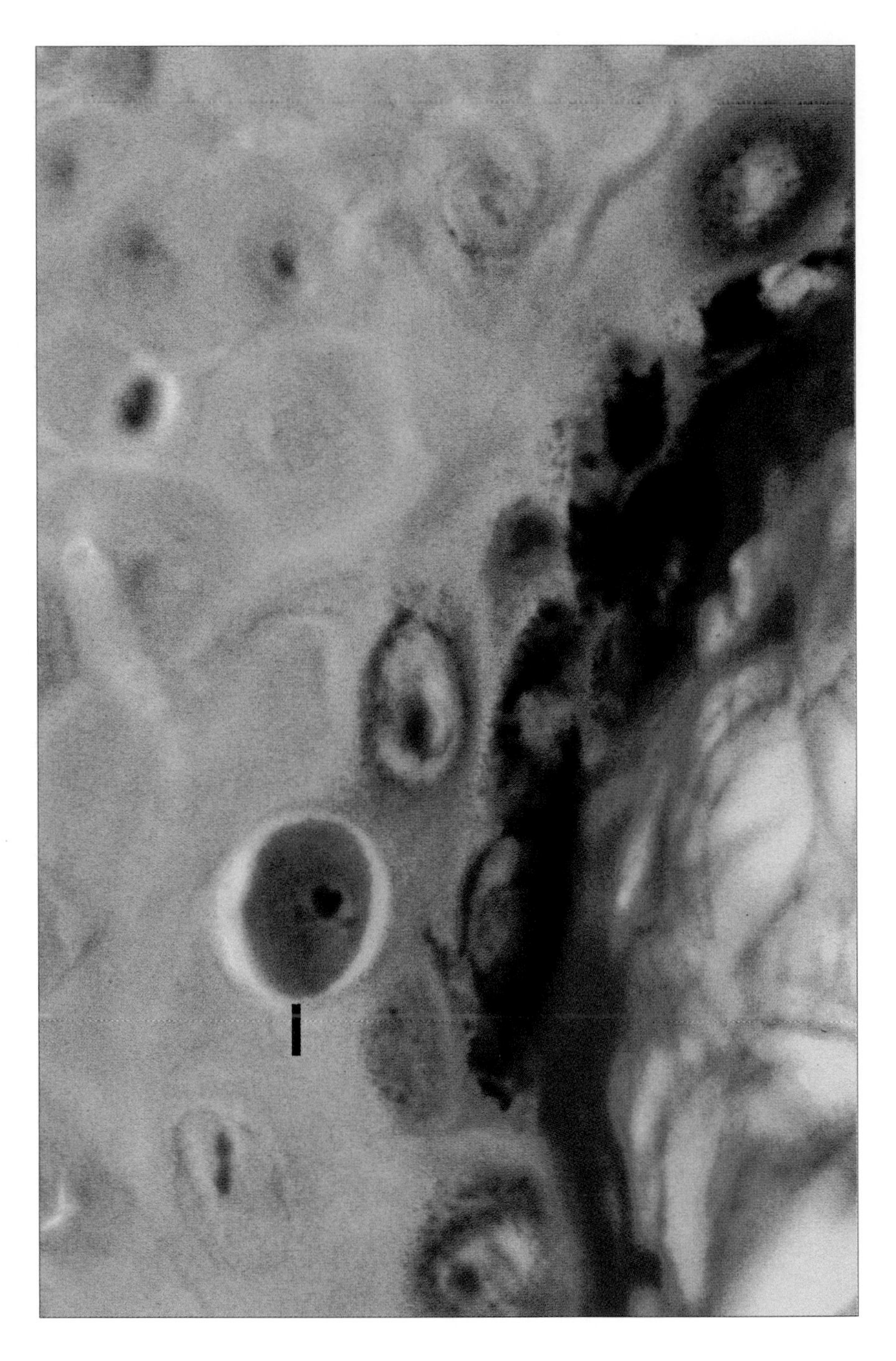

CHAPTER 8: *The cytoplasmic blush of a dying skin cell; other cells do not appear to take note.*

COURTESY OF SCOTT R. GRANTER, M.D.

INTRODUCTION

Choosing a career in medicine was easy; I was enticed by a profession that was concerned for the welfare of others. And contact with patients promised more to me than the ethereal ministrations of the rabbinate. The scholastic intensity of medical school with its required fleeting appearances on hospital wards never dissuaded me, and I was ultimately confronted by a singular choice: What type of doctor did I want to be? This decision is critical, for it defines one's career, lifestyle, family life, modality of service to the sick, remuneration, even prestige. Some of the specialties harbor warrior qualities, and foremost among them is surgery. Back in the sixties when it was my time to choose, I opted for surgery; it seemed to me the other way to go to war.

After I had completed a junior-assistant residency in surgery at the Montreal General Hospital of

McGill University, our program was extended to include a compulsory year of surgical research or surgical pathology. I set out for New York City, Albert Einstein College of Medicine, hoping to familiarize myself with the surgical pathologist's skills for interpreting the biopsies that would be a part of my life as a surgeon. The year promised to widen my medical horizons and introduce me to the microscopic world of cells.

Reflecting back on these two early career choices—to become a surgeon, to do a year of surgical pathology—fills me with amazement: Very little critical thought entered into either decision; they were made on a whim, or so it seems to me now. Never did it occur to me to spend extended time with a surgeon before deciding on this career, to get the *feel* of a surgeon's life. True, I had done my surgical rotations in medical school and internship, but I had barely delved beneath the skin. "The flesh is the spirit thickened," writes Richard Selzer. But I had not yet connected the two.

At Einstein, I slowed to the pathologist's pace. The critical middle-of-the-night telephone calls to convene around a festering belly, and to operate, ceased. I slept peacefully the whole night through and rose in the morning eager to enter the cellular world of biopsies. Here, magnified from forty to four hundred times, were stark cellscapes filled with deep-blue nuclei and scarlet cytoplasm, where cellular appearances and growth patterns seemed to resist my analysis. A devoted patience and steady practice were required to penetrate the kaleidoscopic cellular venues. Coincidentally, it was a provocative time in Einstein's pathology department, for Dr. Alex Novokoff had earlier codiscovered a new intracytoplasmic organelle—the lysosome—and I had exposure to the electron microscopes that he and others were using to magnify cells up to one hundred thousand times or more. In the air of excitement that prevailed, it was easy to replace the patient with the pathology of disease.

After six months at Einstein, I realized a flair for surgical pathology that I had never demonstrated in surgery. For the first time I carefully speculated about my future, and decided not to return to McGill but to complete my training in a specialty I had never once considered alluring during my earlier years of study.

After Einstein, I traveled west to polish my skills at the now defunct Mount Sinai Hospital in Los Angeles and spent a year at the Children's Hospital there. Finally, I took a position at a large community hospital in Torrance, California—Torrance Memorial Medical Center—where I spent the majority of my career.

Throughout my cellular days, which fast became cellular years, surgery remained a part of my experience. And although I have never missed the cutting and tying, the removing or repairing, I have missed the bonding with patients that the intimacy of surgery engenders. I have missed, too, what I still remember of the quiet heroics of everyday people who, in the strange confines of a hospital, struggled in bloody arenas to maintain their dignity. To that end, after so many years, I still seek out patient bonding.

The stories in this book have been written over the last ten years and represent my infrequent excursions into clinical life. They are colored, of course, by my own cellular biases as I attempt to glimpse the lives of patients at different levels.

Paradoxically, the extraordinary people of whom I write have made me see the lives of cells in new and pertinent ways. Human stories can give cellular stories a piquant urgency they cannot otherwise acquire.

THE LANGUAGE OF CELLS

CHAPTER ONE

Working Through Images

My work, as an interpreter of human-tissue biopsies, is largely an art. I carefully observe changes of color, delicately feel for variations in texture, and, with my microscope, peer in on the cells to study their form and tableau. The impact of disease can be very subtle.

The need for my diagnoses to be free of error can provoke unwanted stress. Often the image of a challenging biopsy stays with me for hours, even days. These cells, floating freely in my mind like anxiety, play their tricks, show me their elusive faces, their phantom patterns. They seem to conspire to confuse me. Through the years I've developed tricks of my own—serial sections, step cuts, a host of special tissue stains—designed to counter their deception. When nuclei are marred by craggy clumps of chromatin, and cell patterns appear baroque or abstract, I cull from memory similar compositions and interpretations. After many years at my microscope, the number of different cells and patterns that I recognize, the blueprints of disease, seems infinite. I rely on this experience. And although the majority of biopsies are no longer diagnostic challenges for me, interpretation can, on occasion, be tortuous—but never so formidable as living with the disease itself.

My surgical pathology office is within the hospital histology lab, appended to the surgery suites. A sliding glass window separates me from ten operating rooms. It is twenty-five steps from my desk to that window. A biopsy, if it is to be interpreted during surgery, is processed within minutes of its arrival. I am mindful that the patient is under anesthetic and that time is of the essence.

When I arrive each morning, I scan the operating-room schedule for surgeries with biopsies that will require rapid, frozen-section interpretation. Then I have my coffee in the surgery lounge and listen to the surgeons' stories. A surgeon's demeanor—anxious, diffident, vague—might stir me to anticipate problems, to consult the patient's X rays and chart prior to receiving the biopsy. I am most comfortable with surgeons whose judgment I feel is beyond reproach; they tend to be meticulous, obsessive.

An accomplished surgeon I have practiced with for years tells me about a thirty-five-year-old patient I'll call Hanna Baylan. She has a palpable mass in her left breast; on the mammogram it looked suspicious for malignancy, and the core needle biopsies of it I interpreted a week ago showed infiltrating carcinoma that originated in her breast ducts. This morning she is having a lumpectomy to remove the cancer-containing portion of her left breast as well as a lymph node resection in her left axilla. These nodes are markers for tumor spread beyond the breast. She is worried, the surgeon tells me, that she will not live to see her three small boys grow up.

PREOCCUPIED WITH CANCER CELLS, I have no social or psychological sense of a cancer patient. I retrieve Hanna Baylan's core biopsy slides from the file and review them in my office. I fix on elements of function, not form: milk-producing lobules, milk-transporting ducts, nipples, fat, connective tissue. I fix on cancer. After her surgery, my responsibility will be to classify the cancer, grade its aggressiveness, and determine the extent of its local spread. I will glean the facts that are pertinent to any use of radiation or chemotherapy to help her physicians mount their therapeutic blows.

"Biopsy, room two," the operating-room nurse shouts.

I walk through the histology lab, which smells of formaldehyde. The counters are crowded with vats of tissue-processing chemicals—alcohol, formalin, xylene, paraffin—and glass vessels of vivid red and blue tissue stains. A cryostat—the frozen-section machine standing in the corner—hums like a fluorescent lamp.

Hanna Baylan's lumpectomy tissue, swathed in gauze and labeled, sits on the counter beneath the sliding glass window. With gloved hands I unveil a round fatty mass, its yellow surface smeared with fresh blood. It has the look and consistency of a ripe nectarine. I bisect it with a knife and see a mass the size of a pit at the center, white and gritty as sandstone. Its retracted, deep-rooted look and rock-hard feel imply carcinoma.

The axillary lymph nodes arrive buried in fat. There are twenty-two in all—soft, oval, encapsulated like beans. Two of the beans are hard and white, gritty when cut. The cancer has exceeded its breast of origin. I pass on this information to the surgeon.

At 6:30 the following morning I remove the plastic cover from my microscope and continue my examination. I stare at the sprawl of Hanna Baylan's tumor. The foreboding bulkiness of the cancer cells, the scowl of their thickset nuclear faces looms through the lenses. They are gathered into inane configurations that crudely mimic breast ducts. Although this cancer splays out garishly into adjacent breast tissue, the biopsy margins are free of malignant cells: the local cancer has likely been entirely removed. Eleven of twenty-two axillary lymph nodes bear cancer cells, however, and the probability of spread to other organs is high. I classify this tumor as an infiltrating, moderately differentiated carcinoma arising from breast ducts.

I have completed my evaluation of Hanna Baylan. I await two more breast biopsies, a lung biopsy, and three skin biopsies. All are suspected of being malignant. By tomorrow Hanna Baylan will become a memory for me, a name on yesterday's surgery schedule with tumor cells attached.

By confining myself to cells, I stay clear of the fiery trials of illness. I remain detached; I can render my diagnoses with a cool eye. My fascination with the microscopic form, color, and disposition of cells drives me like a critic to interpret, to applaud or decry them for others. Paradoxically, observing so much of life through a microscope all these years has left me feeling, lately, that I've sampled too little, that I've missed the very warp and woof of it.

"Dr. Nadler?"

A young woman is standing at my office door.

"Sorry if I'm disturbing you, but no one was at the reception desk so I walked right in," she says. "I wonder if I can see the slides from my breast tumor?"

"Now?" It's six o'clock, the end of a long day.

She enters and sits in the chair by my desk. "You don't remember me, do you, Doctor?" she says. "I was at the lecture you gave at the Wellness Community last month."

Her cropped blond hair has a uniform thinness that suggests chemotherapy; her face is gaunt and pale. Still, she seems valiant somehow, undaunted, her self-esteem intact. During the lecture I had used a projecting microscope to show on-screen what the cells and patterns of different tumors look like.

"I'm Hanna Baylan. You diagnosed my cancer forty-three days ago."

I don't recall seeing her at the lecture, but I do remember, in vivid detail, the nectarine lineaments of her lumpectomy tissue. I'm like the surgeon who selectively focuses on the organs he's rectified or removed. My work lies apart from Hanna's face, among the tiniest kernels of bodily things; my work lies in her cells.

"It's pretty late," I tell her.

"Yes, it is," she says. "Maybe it's already spread to my bones."

This is not what I meant. "Why don't I see what I can do." I wish she had called ahead, given me a chance to review her slides.

I retrieve all her breast and lymph-node slides from the file and move her chair opposite mine. With effort, pain maybe, she leans across the desk to peer through the alternate set of eyepieces on my two-headed microscope. Resting her elbows on the desktop, she looks in on the events of her body—cells long dead, now fixed and colored—that have given rise to her illness.

She listens quietly as I move the pointer across the microscopic landscape. "These clustered islands of glands are the lobules," I tell her. "Milk is produced here in the lactating breast."

"They look more like pink hydrangeas to me," she says, "a sprawling garden of them." She talks excitedly, asserting interpretive authority over her own cells. I can only imagine the variety of forms a cellular array such as this might suggest to an uninitiated eye.

"And these?" she asks. "What are they?"

"Ducts," I say. "They transport lobular milk outward to the nipple."

"My God. Look at them," she says. "Ponds, lakes, rivers, estuaries that carry milk. It all looks so peaceful." With her legs braced in the chair, she hoists her torso onto my desk and hunches over the microscope to get a better view.

There is little need for pedagogy; she is finding her own truths with metaphor. I switch from the four-power objective, the scanner (a magnification of forty), to the forty-power objective (a magnification of four hundred), and individual cells take prominence over cell patterns.

She clasps her hands together. "It's as if all the planets in the universe have come together here."

“See the uniform cells lining the lobules and ducts?” I point out the blue nuclei, the pink cytoplasm, the discrete nuclear membranes.

I switch back to the scanner and we pass over fields of ducts and lobules. I suspect it is a whimsical leitmotif she sees, hydrangeas, ponds, rivers suspended idly in fat and fiber, floating serenely.

I wait a few minutes, allowing her to absorb the beauty of her own cells. Sitting perfectly still, crooked like a pliant ballerina, she inspects her cellular self.

Reluctantly, I replace the slide of her normal breast tissue with one of the cancer.

“Whoa,” Hanna says.

She stares into the microscope, transfixed by the disarray of her malignant growth, a raw view of her life spread out before her. “These cells look like distorted hula hoops,” she says. “It’s all damaged, isn’t it? Just like my real world.”

“This is your real world, too,” I say.

She looks at me over the top of the microscope. “People don’t shun me because my tumor ducts look like reckless hula hoops.”

Like Charon ferrying between the living and the dead, she glides back and forth between her threatened life and her dead, stained biopsy cells. She quickly grasps the cause and effect—critical cell changes are twisting her life. For years I have processed thousands of such cases, determined the manifold forms of disease, but I've never been an intimate part of anyone's illness, never felt the connection of cells to a larger self.

"Losing my hair terrifies me," she says. She fingers it, pulls at it gently. Not a single strand comes out, and she is reassured. "I've got a wig, but I hate it. So I wear baseball caps and tie scarves through my hair. I'm lucky. I look good in scarves. Still, I feel hideous. People think it's just vanity. It's much more than that," she says. "Every time I see my scalp poking through, I'm *reminded.* I feel how different I am. How lonely."

"You'll have your hair back in a few months," I say.

Tears well in her eyes. "That'll help a whole hell of a lot."

In *The Notebooks of Malte Laurids Brigge,* Rilke writes, "If I am changing, then . . . I am no longer the person I was, and if I'm something else, then . . . I have no acquaintances." I believe that Hanna's perception of her disease-tainted self is one source of her loneliness; she will have to be her first new acquaintance before others can come along. And it pains me that all I can offer her is my familiarity with her cancer cells.

"What chemotherapy does to me is unbelievable," she continues. "After a treatment I wake up around midnight with a funny taste in my mouth, and then boom, an incredible indigestion—like a volcano—with nausea and vomiting that rips my insides out. It's excruciating. Every bone in my body aches. Things stop for a while, then it starts all over again. Off and on for the rest of the night."

She is on Cytoxan and Adriamycin, she tells me. These drugs act during mitosis to prevent cell reproduction, destroying the rapidly growing cancer cells, hair cells, bone marrow cells, and cells lining the gastrointestinal tract. Hence the tumor destruction, hair loss, reduction of blood cells, nausea, and vomiting. It's a savage exposure, a supervised chemical warfare.

"I was alone in bed one night last week," she says, looking up from the microscope. "My husband was out of town, my kids were asleep, it was after midnight. I lay there staring at the ceiling, scared out of my wits, shaking uncontrollably. Suddenly a warm white light beamed through the window and rested on my chest. It was a miracle, the way it soothed me to sleep." She slides back into her seat. "I realized when I woke up that God was looking out for me."

I am moved by the way Hanna aligns herself with all her positive expectations.

SIX YEARS GO by before Hanna Baylan reenters my life. I have not asked after her, nor have I received word of her struggle. I have retained the professional cool, the isolation that has been so much a part of my life.

Once again Hanna appears at the end of a long day. She walks slowly and sits down with some difficulty in the chair by my desk. She's frailer now, and her pallor makes her eyes seem dark, watchful. Her cancer, she tells me, recurred the year before. Three spots in her ribs, one in a lung. She submitted to high doses of chemotherapy, more toxic than before, then underwent a bone-marrow transplant.

"The cancer in my bones was like a little old lady," she says. "It puttered around, came and went. But I could deal with it." Her jaws tighten. "It's the drugs, not the cancer, that are so hard to take. People who haven't had chemo never really understand that.

And it's the fear that you may die. It's been hard to come to terms with that."

She outlasted the poisons, metabolized them. The cancer in her bones and lung disappeared from view. Cells harvested from her marrow before this chemotherapy were then returned to replace what the drugs had destroyed, hopefully to spawn a new remission.

If she is to succumb to her illness, her bearing shows no hint of defeat.

"I'm here to see my cancer cells again," she says. "I'd like to see them projected on the big screen, like you did at your lecture." Her arched brows reflect her resolve. "I need to confront them one at a time to get a handle on them."

I set up the xenon projector in the hospital auditorium. Before long we are alone in a large, quiet space.

I project one of her biopsy slides onto the screen, magnifying her cancer cells to the size of golf balls. They glare at us like cyclopean monsters—granular pink bodies clinging to one another, each nuclear blue eye reflecting its own confusion.

She walks slowly down the center aisle, passing

orderly rows of chairs. The large screen hangs on the front wall between a varnished wood lectern and a row of X-ray view boxes. "That's perfect," she says. "I want to see these guys up close and personal." She touches the screen, runs her fingers over her cancer cells as though gathering their random spread into some kind of coherence. The loveliness of cells on slides, all the different shapes and colors, allows Hanna to give her breast cancer its own identity.

"They're like moons," she says, "each with a different face, a different complement of light and dark."

I turn off the auditorium lights. Her hands spread a silhouette that shadows her moons like eclipses.

"I'd like to spend time here," she says. "Touch them, get to know them."

"No rush," I hear myself say.

I stand at the projector in darkness. She is forty-one, as I recall. Her mother died of breast cancer in the 1950s, when it was considered a local disease. The ideal treatment back then was to resect as much local tissue as possible—the entire breast, underlying pectoral muscle, lymph nodes—before the cancer could spread. I recall receiving so many of these horrific specimens in the 1960s that I felt like Artemis of Ephesus, the Great Mother of Life, whose torso teemed with breasts. It was disquieting to conjure up an image of the women attached, only moments earlier, to the specimens before me. We now know that breast cancer is not necessarily a local disease, that it can be a covert presence in the breast for years, shedding cells that rain through blood and lymph vessels and sprawl to other body parts long before the primary mass is discovered. Hanna's cancer cells had obviously left home, finding sanctuary in her axillary lymph nodes, by the time she felt her mass. She needed drug treatments then to defend her body from the spread. Although her breast surgery was far less aggressive and disfiguring than her mother's, Hanna's drug therapy has greatly exacerbated her difficulties.

"I'm returning from Lilliput," Hanna announces. She walks back along the aisle, spotlighted by the projector beam, looking as hopeful as a bride.

When I turn on the lights, she is standing beside me. "I must confide a strange thing," she says. She slouches, her body seemingly depleted by the encounter with her cells. "When I finished my last cycle of chemo, I felt no sense of relief. The thought that a few bad cells could be hanging around, and that nothing more was being done chemically scared me. It still does."

I feel the uneasy edge between her confidence and her fear. Others are better suited to help. But I realize that I can listen.

I HAVE NEVER UNDERSTOOD the purpose of a newspaper obituary. As a published notice of death, it certainly works well enough. As a biography filled with concrete facts—achievements, mostly—it gives the life in question a one-sided loftiness devoid of the flaws and failures that make it whole. And where is the mention of an individual's spirit, his effectiveness as a human being, his courage in adversity? What about people who successfully battle illness for many years before they succumb? What are their achievements in this regard, or do they simply "die after a long illness"?

Hanna Baylan's illness is very long. Four years go by before she comes to see me again; she has more questions about her cells.

"I must be stupid. The cancer is back, spread to my liver, but I just don't get it that I'm supposed to die," she says as she enters my office.

I notice how the ridges beneath her eyes have darkened, discolored by years of anguish and fear. She grimaces as she moves, pain loose inside her, and settles cautiously into the leather chair in the corner beneath my bookshelves. A CADD pump is fastened to her waist, a beeper-sized gadget that pumps chemo into her subclavian vein.

"I want to know more about these little buggers inside me, what really makes them tick," she says.

I assume it is the internals of cancer cells, the organelles, that she refers to. They are too small to be seen effectively with my microscope, so I show her some black-and-white electron micrographs of cells magnified up to one hundred thousand times. She stares at the oblong mitochondria whose cristae resemble zebra stripes, and at the round, secretory vacuoles that look as dark and heavy as medicine balls. I find her a freeze-etched micrograph of a nucleus that truly resembles the desolate, pocked surface of the moon.

She studies the micrographs, keeps an inquisitive silence. I await the new metaphors she'll conceive to keep her cancer at bay. I am thinking how our imagination is what saves us.

"Why do cancer cells keep growing and multiplying if they're so destructive to the body?" she finally says. "Why don't they just die?"

She's tired of all the pretty pictures, the metaphors. She's ready to deal with her cancer in a more direct way. I tell her that our dysfunctional and superfluous cells normally self-destruct in a programmed cellular suicide.

"Don't cancer cells self-destruct?"

"Apparently not. Cancer somehow disables the program. The cells forget how to die."

"Well, so do I," she says, a faint smile stealing across her face.

Then suddenly she starts to cry, shaking as though grief has surfaced from all the deepest places. "I've been blessed with three wonderful boys and a husband who loves me," she says. "They'll be devastated if I go . . . so I can't give up."

I put my arms around her, give her a long, firm hug. Her bones seem ungraspable, like hope.

I BEGIN TO SEE that the diagnosis of a disease plays little part in the healing process; so, too, for that matter, does the treatment strategy. Help attuned to individual needs is what heals. Disease seems to be more than a set of facts, and illness more than a diminished way of life; they are a strange tandem that plays out differently in every host: despair, terror, agony, a call to arms, newfound clarity, transcendence, metamorphosis. Those afflicted must have their needs satisfied on *their* terms. *They* must control, as much as possible, the progression of their own adversity. I can feel Hanna yearn for answers. I must give them to her, show her the pictures that help her.

And yet I feel a separation from Hanna, how her cancer intangibly intervenes; I must be prudent to be effective with her. This is heartrending to me, for I have come to love her—her forthright style, her spirit, her desperate determination to nurture her family as long as she can. I can no longer think of Hanna in terms of the cells I see on her slides.

She leaves my office, and I feel the loose strings in my life tighten.

A FEW MONTHS PASS and Hanna returns with her youngest son. He has decided to be a doctor and she wants to introduce him to me. He has his mother's poise, her small, delicate features, and he fixes me with probing eyes.

"Dr. Nadler shows me his biopsy pictures, lets me observe my disease," Hanna says, "so I can see what I'm up against." She is barely able to contain her pride as she sits beside her son.

She's fought her cancer to a stalemate. It's become more like a chronic illness than a life threat. And her son has lived so long with his mother's cancer that it is much more real to him than the likelihood of any cure.

The flesh of Hanna's successful life sits beside me. I answer all his questions, try to inform his career decision. I think he knows how valiantly his mother has girded herself on his behalf.

"I feel another flare-up coming on," Hanna says. "I'm strangely tired, not quite right. So I've decided to go to Maine."

"What's in Maine?" I ask.

"Leaves," she says. "I'm going to see the fall."

I picture metastasizing tumor cells pushing their way into her lungs and liver and bones. Can she muster her immune system once again? Will the drugs still have potency for her? Will there be another beginning?

When Hanna Baylan and her son leave my office, arm in arm, there is a confidence about her that seems complete. It's as if she knows that falling is followed by rising, and that eternal falling leads to a rising in another time and place.

I return to my microscope. In the scarlet spread of a squamous skin cancer, I strain to see the deciduous autumn leaves of Maine, so fiery when first fallen, then turning slowly to compost, to nurture blanketed seeds.

CHAPTER TWO

Fat

UNLIKE CANCER CELLS, FAT CELLS ARE so often part of a biopsy's cellular mix that they can easily be taken for granted, glossed over as I seek more likely cellular culprits in the structuring of a disease. The aggregate size of fat cells, however, can affect our health, shape the way we think about ourselves, and how others think of us. It can even make our lives unbearable.

Patti Fleming, at five feet, eight inches tall, once weighed three hundred and fifty-six pounds. Now less than half her former weight, her face is pretty and her body trim. Although her colossal subtraction yields nary a physical clue of her former self, she tells me that in her mind she remains a fat person.

Patti is reluctant to come to my hospital (any hospital) to talk with me, so I take the afternoon off to visit with her. I am not averse to searching out patients' stories on their own turf where they are often more forthcoming than on mine. A medical colleague who knows of my interest in breaking through a surgical pathologists' ingrained clinical myopia has introduced us and we sit in a sterile conference room in the law office where she works as a legal secretary.

She confides to me how implausible it was that she was ever so obese. Her maternal grandmother and her mother "carried weight in their stomachs" but neither so extremely. Patti was distressed by her metamorphosis, by the constraints, hazards, shame, and fragility of it; she was also concerned by the adverse effects it had on her family.

"I first noticed that I carried too much weight when I was ten years old," Patti says. "I couldn't get up on a horse as easily as the other girls."

At camp at the time, she ate the same portions of food as the others, exercised every bit as much. Were "fat genes" beginning to express themselves?

"In high school, other kids first talked about me being fat, but I wouldn't let it happen. I'd skip meals and was very active in sports. I knew that if I ate, I'd get fat." She laughs. "I felt like if I smelled a cake, I'd gain five pounds."

I sense the loneliness, even helplessness, as Patti Fleming tells of beginning to discover, early on, her fat cells' penchant for hoarding fat. Her parents, away at work, didn't know that she seldom ate breakfast or lunch. She had her dinners at home and her parents didn't suspect her dietary dilemmas. On weekends she was out of the house, skipping meals and trying to ignore her hunger. Her private starvation was a mind game, as impermanent as satiation.

When she met Frank Fleming, her husband-to-be, he was unconcerned about her weight. He encouraged her not to skip meals.

"We enjoyed eating together. I just love food. We ate the same amounts and he stayed skinny . . . I got bigger and bigger."

They overate. Despite studies that suggest altered metabolism or disturbed satiety signals as predisposing factors, one must overeat to realize morbid obesity (one hundred pounds or more above the ideal body weight).

Whenever Frank gained a few pounds, he stopped eating sweets for a week or two, and fat seemed to melt from him. From the end of her teenage years until she was forty, Patti "pushed and shoved" her way through dozens of diets, spending thousands of dollars, losing and regaining hundreds of pounds. She tethered herself to diet food until she could no longer tolerate the tedium. The priceyness of those special packages angered her; society did, too, for the stigma it placed on her body.

"I tried to exercise, but there are no commercial gyms that can hold people who are more than two hundred and fifty pounds. You break their machines and they want you out of there."

At her peak weight, exercise did not come easily. It was a major effort for Patti to walk down the street. "My heart would pound, my mind would race, and I couldn't stop sweating," she says. "I finally got so big that I was incapable of exercising."

Her fat-cell aggregate was now so huge that she could barely keep her head above it. She seemed to be sinking inside herself, too mired to surface. Rather than garnering sympathy, she often repelled people; it was as if her globate habitus revealed all that was inside her, negating the need to look into her eyes or listen to a single word she spoke. It is hard for the morbidly obese to stand up for their rights when others consider their size and shape a character flaw rather than a disease or disability. "Stop stuffing yourself," people say, ignorant of the power of fat cells. "Do a little exercise once in a while." "Don't be so damned lazy."

"I'd sleep on my side because I couldn't breathe on my back, and my arms and hands would get caught under my body, cutting off the circulation and waking me," Patti tells me. By morning, her upper limbs were numb and swollen, and her back and neck were sore from all the awkward posturing. The mattress on her side was cratered like a sinkhole. When she rose from it, she covered this depression with a blanket so Frank could not see. This was one of the many oppressing visuals of her disease that she strove to hide.

"I'd clean my private parts with a hand shower first thing. Morbidly obese people won't tell you this, but they can't wipe themselves properly after they use the bathroom. They can't reach their tush." She has no shame now. There is a palpable sense of pride in having taken her fat cells to task, gotten them to release so much of their fat. "I kept a bottle washer with me during the day, and I'd wipe myself with it the best

I could if I needed the bathroom, but how uncomfortable was that? If I was sick, I had to rely on Frank to wipe me. He never said a word, but I was humiliated." She talks openly about all that is personal so that others may come to know how restricted morbidly obese people are, why so many of them would rather be deaf or blind than suffer the affliction they have. "You're hardly alive when you're so huge. All the time I'd think about my kids, how embarrassed they were for me, how I couldn't really be there for them. I'd have given anything to lose weight permanently, but I just couldn't do it."

Beneath the massive bulge of abdominal fat, ulcerated skin rashes wept until the smell was so rank she could not stand it. And none of the medications she used brought relief. It itched and pained her until she was almost mad with suffering. At home, she raised her abdomen and sat in the sway of a fan until the forced air dried her wounds, and she had temporary relief.

I ask Patti to recall a typical day with her largely atypical body.

"I never ate breakfast," she says. "I got up, showered, and drove to work. The car seat creaked and crumbled beneath me. I was hardly able to reach the steering wheel with my hands for my belly, and my feet went numb from body compression."

"Once I was rear-ended and broke my seat in half. Another time I hit a car, broke my seat belt, and flew through the windshield." With such great weight comes a staggering momentum.

There was little discrimination in the workplace, but lunches were different. She worried whether she would fit into someone's car if they all went out to lunch, whether she would be able to wedge into the restaurant booth (she prayed for tables and chairs to save humiliation). Maître d's and waitresses ignored her. *Does anyone her size really want more food?* And sitting so close to others, she worried that the smell of her belly rashes would be detectable.

By the afternoon, Patti would feel emotionally and physically exhausted. Her legs swelled from supporting her body and sciatica pierced her lower back like a fiery poker. When she finally got home, she was too toilworn to cook, so she bought fast food for dinner. It was easy and cheap, and Frank and the kids liked it.

Some nights, the parent meetings at the school could go on for hours. "The chairs they had in the auditorium were little plastic things. I'd stand, but eventually they'd say 'sit.' " So she would spread herself over the chair in a very cautious way, her legs and thighs bearing much of her load while she braced her arms against the armrests. With her body distributed in this way, her massive bottom rested lightly upon the seat. And there she "sat" for hours.

At night she was exhausted. If she and her husband had sex, there was no joy in it. She was confined to the missionary position and could not remain there long before she began to choke.

Her lumbering lack of mobility in bed was yet another struggle, another way in which she would fail to perform. Her ineptnesses were born of severe restrictions and often magnified by the callousness and ignorance of others. Beaten daily into feelings of inferiority, she finally concluded that it did not matter how fat she was. She had no hope.

OF ALL THE CELLS I see, few are as distinctive as the human fat cell: Inside itself, a large fat globule steamrolls other cell contents flat against the outer membrane until it bulges like a mozzarella. Freeze and section this cell through its nucleus and you see a signet ring snugly fit upon a fatty finger. With the scanning electron microscope, there is such a loss of fat in the processing that what remains is a three-dimensional shell of this cell's former self, cytoplasm and cell membrane stretched into a rim.

In all its endeavors, the fat cell is buttressed by an external network of collagen fibers, where myriad capillaries, even occasional nerve twigs, course in between. By and large gregarious, fat cells gather into millions of lobules; these are separate from one another yet held like a vital mosaic by miles of inflexible fibrous tracts.

Fat cells teem inside us: They pervade the great fat layer beneath our skin, congregate around our adrenal glands and kidneys, in our abdomen and chest and bone marrow, in the grooves of our heart and the subtle spaces of our neck and armpits and groin. Our fat cells are everywhere, save the central nervous system, lungs, eyelids, ears, penis, and the backs of our hands. They insulate, buffer, and energize us.

The average-size infant enters this world with approximately five billion fat cells, one sixth to one seventh of the adult quota. In the ensuing six months these cells significantly sufflate with fat, but their numbers remain relatively constant. From then on to the end of puberty, it is the number of fat cells that increases to adult proportion, their size unchanging. And throughout adulthood we fatten mostly by adding to the fat in our fat cells; some of us are capable of distending them by a factor of three or four to a gargantuan three-hundred-micron diameter. Fat cell distention can also lead to *de novo* fat-cell formation, perpetrating a still greater mass of fat. If our thirty-five billion adult fat cells enlarge in these ways, we can balloon to rotundity and come to resemble the sphere of our fat cells.

In human biopsies, normal fat is a glistening, uniform cadmium-yellow, its texture greasy and soft. In vivo, at body temperature, fat is liquid, oozing in and out of fat cells under neural and hormonal control. There is a perpetual ebb and flow of fat between its mobilized state in the blood (free fatty acids) and its storage state in fat cells (triglycerides, mostly). The free fatty acids burn metabolically as a high-potency energy source, and what is unused by the body gets restored as triglycerides in the fat cells. Such is the dynamism of fat, a flux complexly modified by how we eat and exercise, as well as by our genes.

Envision this giant fat "organ" tucked within our bodies, how it can rise to roll our features outward or shrink until it fades among the splay of protruding bones. To make sense of this organ when it is a lipid-laden albatross, we must hark back to our distant hunter-gatherer ancestors. No fatness there. It has been theorized that these ancestors acquired "thrifty genes" to store the fat of feasts, to sustain them through the famines. In today's American surfeit, where feasts are the norm, these ancestral adaptations have become liabilities. Fine-tuned by our individual genetic legacies, each of us settles into a metabolic equilibrium. For some, this settling point renders us obese, and it is difficult to reduce this fat without a sustained dietary effort. And the increasing consumption of cheap, readily available fatty foods is propelling the most metabolically susceptible of us into exorbitant obesity. Once our fat cells become extremely impacted, the failure rate of diets is almost 100 percent. For four million morbidly obese Americans, diets strict enough to succeed are usually too severe to be feasible. The unfitting long-term results of medical, drug, and behavioral therapies for obesity in extremis have increased the number of referrals for bariatric (obesity) surgery. An illness without a cure, complicated by heart disease, hypertension, diabetes mellitus, sleep apnea, gallstones, degenerative arthritis of weight-bearing joints, restrictive lung disease . . . morbid obesity is spreading among us. We are coming together into lobules.

Sometimes the opposite can happen. Markedly reduced or absent fat is usually a marker of serious illness, a sign of the malady itself. Most often one loses weight occultly with cancer and overtly when a lack of appetite springs from an obvious underlying physical or psychological ailment, or starvation from famine or subjugation. Seriously underweight people appear hollowed out, as if their flesh has sunk beneath their skeletons, and their bones, flagrantly protruding, get locked together in caricature. It is sadness we feel for those who have no fat, and anger if this loss is engendered by tyranny.

Depleted of their fat, starving fat cells shrink until they are rounded to fifteen-micron-diameter versions of themselves. With centrally placed nuclei and globule-free pink cytoplasm, they come to resemble tumor cells. And brown-gold lipofuscin pigment granules (footprints of wear and tear) lightly disperse themselves throughout. As part of this involution, fat lobules can deflate into discrete, fat-free balls or, in the severest cachexia, into wormlike streaks, wisps, as they distance themselves from one another. This fat loss is most apparent beneath the skin and in fat that hovers inside the abdomen.

When starvation remains unabated and the body's fat stores are eventually spent, proteins are burned to fuel the last flickering glow of life force. But proteins are essential for maintaining cell function. And death ensues when proteins are depleted to half their normal level.

SURGEONS FIX BODILY THINGS. What is removable, if diseased or malfunctioning, they can remove, and what is irremovable, they can sometimes imaginatively bypass. Today surgeons are circumnavigating portions of the stomach and small intestines to enable the morbidly obese to cope with sustained weight loss. Patti Fleming, with the help of her doctors, comes to believe that bariatric surgery will retrieve her earlier life, that without it she will not likely live to see her grandchildren. It is a courageous endeavor to submit one's obese body to the knife in this way, to pucker billions of fat cells in the hope of renewal.

The size and structure of the stomach pouch, the nature of the intestinal rearrangement, are intimately related and the subject of much surgical bandying. One such gastrointestinal reconstruction—the Roux-en-Y gastric bypass—is currently the procedure of choice at UCLA; it is the one that Patti Fleming agrees to undergo. The bariatric team is convinced that she is a good candidate for the procedure, that she can commit herself to a draconian modification of her gastrointestinal tract, and her life.

None of Patti's family or friends encourage her in this surgical pursuit. Surely another diet is preferable to the knife, they argue; perhaps this time she can prevail against her greedy fat cells. Although Patti has been dieting unsuccessfully for years, her loved ones seem to deny what is clearly evident: She needs additional help to counteract her "thrifty genes," the settling point of her metabolic equilibrium.

She awakes from the surgery frightened by what has been permanently perpetrated. Her eating habits will forever be altered and she must permanently commit to supplementing her new diet with vitamins that are no longer adequately absorbed by her new designer gut.

Picture Patti's 1,700-milliliter stomach reduced to 35 milliliters, miniaturized. This drastically reduces her tiny new stomach's capacity to process, absorb, and propel foods. And her small intestines are now reconstructed to reduce the number of calories they can absorb. Patti has to limit her food intake to frequent, small-portioned meals or suffer the consequences: The trim stomach pouch, unable to handle at one sitting a food serving larger than a hard-boiled egg, or one that has been chewed fewer than thirty or forty times, will dump its unprocessed contents into the adjacent small bowel, causing abdominal pain, nausea, diarrhea, dizziness, heart palpitations, even loss of consciousness. Experiencing this "dumping syndrome" only once, Patti is no longer inclined to swallow other than bonsaied boluses at each sitting, and takes great care to chew them to a flow.

The postoperative abdominal cramping is so severe that it makes the deliveries of her children, without anesthetic, seem like child's play. Support groups that now are commonplace before and after bariatric surgery have not yet been organized, so she endures her flesh wounds as she has her obesity—largely on her own. Though Frank has not encouraged her to undergo the surgery, he quickly accepts her decision and helps her to heal.

From the smallest drinks of water, Patti progresses to clear fluids and sugar-free Jell-O. By the third week, the wound pain and cramping subside, and she swallows liquid meal replacements (Carnation Instant Breakfast, Slim-Fast). By the sixth week she is consuming high-protein, pureed foods and beginning to eat select soft foods (nonfat cottage cheese, water-packed canned tuna). By the eighth week she is eating fruits and vegetables, adjusting to the small-portion, low-fat diet that she will live with for life.

Fat cells are releasing their stored triglycerides, and free fatty acids in the bloodstream are burning like oil. Fat is melting into energy and Patti feels her clothes coming loose. She is still over three hundred pounds, but Frank notices the shrinkage and cheers her on. She feels the urge to exercise and swims in the apartment complex's outdoor pool at night so the neighbors cannot see her. And pangs of withdrawal from fats and sweets seem to wrack every shriveling fat cell. She cannot even drink a soda; the carbonated bubbles take up too much space.

As pounds are shed, Patti's high blood pressure returns to normal and the arthritic pains in her feet, knees, back, and neck subside. She is no longer primed for illness.

"I lost so much weight in the early stages," Patti says, "that I didn't know what size I was. But it took my breath away when I first went to Victoria's Secret to buy underwear, and to Robinson's-May for my clothes. I never went back to Lane Bryant again."

Eating is no longer the big pleasure thing, the one enjoyment it used to be. Patti still enjoys food, but it is a slow and cautious satisfaction. She can eat many of the foods she wants as long as the portions, the fatty foods, and refined sugars are curtailed and carefully chewed.

Patti's surgery is not so much a solution as a tool. It allows her to steadily deplete her fat cells, to sustain this depletion so long as she plays by the rules. "If I get stressed and forget to eat slowly," she says, "the food gets stuck. It's like an elephant walking on my sternum. Frank will quickly spot fast eating and hold my hand. It's one of our secret little communications. He knows how hard I've worked to get where I am."

A state of euphoria slowly builds. It is six months after surgery, and Patti feels as though she is but a core of her former self. Her jaw, elbows, shoulders, breasts, buttocks, knees, even her lap emerge from her dwindling sphere. She sits in a regular chair without bruising her hips and easily crosses her legs. In restaurants, she "falls" into a booth, and she effortlessly climbs in and out of cars. She can easily manipulate exercise equipment and get into elevators without a second thought.

The lighter she gets, the greater her incentive to retrieve all that she has missed. In her new one-hundred-and-sixty-nine-pound body—she settles there—Patti Fleming does just that. Without the specter of her mammothness, she interacts with people and things in casual, stress-free ways. But it is her body's feathery lightness, the mobility of its trimness, that most excites her. She camps, water-skis, and kayaks, as she did in her youth. She wears tight-fitting dresses. Patti has her life back.

ON THE STAGE of my microscope, fat cells, whether too large or too small, aggregate in a lobular, geometric precision. Theirs is a glomerate beauty, an abundance or dearth of pure energy. In this microcosm, I can see the power of their mutability. No other human cells can so rampantly rise or fall and, like millions of fiery suns burning bright or burning out, alter our universe.

CHAPTER THREE

Brain-Cell Memories

THE BRAIN, UNLIKE THE RHYTHMICALLY contracting heart, sits motionless in its cranium, no more animated than a liver or spleen. Roughly the size and weight of a cantaloupe, it has the uniform consistency of cream cheese. The gray exterior undulates like a bust by Giacometti, and the homogeneous white interior conceals its daedal scope.

Under the microscope, however, brain cells come alive. Stained with a gold or silver impregnation, these blackened neurons dazzle with their inimitable forms. I've seen them sprout wispy tendrils so long as to seem boundless, coursing the brain like so many fault lines. Neurons can simulate crabs and spiders, brambles, even ornate chandeliers. The smaller companion glial cells, though less conspicuous, are also branched; these groundskeepers outnumber the neurons by ten to one and fill in the spaces between them.

Envision your brain, its billions of impulsive neurons, tendrils entwined, connected up in electrical circuits, elaborate glial scaffolds shoring up these circuits like electrical tape. In the fluid-filled clefts that separate neurons from one another, electrical impulses convert to chemical ones. Brain chemicals seep through the cleft fluid, bridging it with a flow of molecules. Thus do we smile or weep, plumb the ocean or fiddle a tune.

I ANTICIPATE a brain biopsy cautiously. Microscope in hand, I peer in at the musters of neurons and glial cells, taken alive, caught unaware in the course of biopsies. These microcosms never appear morbid, for the cells, frozen in the midst of life, have lucid immediacy. Although technically they are dead, their images challenge me to think about their lives: I am familiar with neuronal changes wrought by disease but know little of the responsibilities each brain neuron bears for thought, emotion, action.

On my desk this morning are biopsy slides from a large tumor, born in the brain of M.K., a sixty-three-year-old Caucasian male. To me, this man is a name-with-tumor-cells-attached, not a frightened human being in the throes of his harrowing diagnosis. His tumor is derived from his glial cells, a glioblastoma multiforme. The euphony of the name belies its malevolence; it is the most common and malignant of glial tumors. He presents with disorientation, incontinence, and progressive left-sided weakness. The tumor, viewed on brain scans, fills a large portion of the right brain and extends through the body of the corpus callosum (the conduit of nerve tendrils between hemispheres) into the left brain. This image of a central body of tumor unfurling laterally into both hemispheres is a common advanced glioblastoma growth pattern likened to the *Lepidoptera* or butterfly. A pterosaur—perhaps the dragonlike *Dsungaripterus,* a flying reptile far less alluring than the butterfly—seems more appropriate.

M.K.'s tumor, examined in the operating room, is too large and entangled with normal brain tissue for the neurosurgeon to remove; he submits a biopsy wrapped in gauze to the surgical pathology suite. The tissue I receive is variegated—the reds and browns of hemorrhage, the yellows of tissue death (necrosis). Under the microscope, the tumor cells vary from unobtrusively small to grotesquely large. Hallmarks that enable me to diagnose glioblastoma multiforme are the exuberant "grape bunch" proliferations of small vessels and the jumbled palisade of tumor nuclei that gird necrotic tumor patches. I've seen these cellular glyphs time and time again in glioblastomas.

Although the presence of this tumor saddens me, mine is a dispassionate sadness, a certitude that the microscopic events I see will soon culminate in M.K.'s demise. If he is to fight against all odds for his survival, I will not be privy to his gallantry. Although I may learn from his diligent neurosurgeon how long he survives, the tumor itself, the maleficent pterosaur, is my only understanding of this man, my only connection to him.

To fathom the formidable malignancy of glioblastoma multiforme, one must go beyond the two-dimensional pictures of microscope slides and conceptualize the disease in four dimensions: the three spatial vectors and time. Think of a glioblastomatous pterosaur, its malignant cells spreading outward in all directions into the surrounding normal brain, creeping, in the bungling way such cells do, along neuron tracts, arbitrarily coursing through circuits until they short them. The complexity of such a geometry, the numbers and locations and functions of shorted and soon-to-be-shorted circuits, is incalculable, the malignant glial progression all but unstoppable, the patient's behavior and loss of function largely unpredictable. Think of a shot put propelled through the consecutive strings of a million tennis rackets; this conjures the enormity of the glioblastoma's destructiveness.

In his essay "On Probability and Possibility," Lewis Thomas concludes that "we pass thoughts around, from mind to mind, so compulsively and with such speed that the brains of mankind often appear, functionally, to be undergoing fusion." I am aware of this fusion when I am captivated by meditations, poems, or trenchant discourses. My microscopic vision of a brain biopsy creates a different fusion: when I see the precarious state of M.K.'s neurons in the wake of his spreading tumor, they seem to call out to my own, an urgent message to the living from the cellular dying and the dead.

In his eloquent memoir, *Death Be Not Proud,*

John Gunther recounts the life of his son Johnny, and his death from glioblastoma multiforme. The tumor lingered after each therapy as if licking its wounds and, quickly rebuilding its autonomous self, began again to challenge the boy's brain for the fixed space in his skull. Gunther details how his son maintained his faculties—intellect, charm, ambition, courage—despite a fifteen-month deterioration. It is an account that testifies to the glioblastoma's willfulness. Once its blundering cells reach the crawl spaces between neurons, it clings to its host like a weasel. Although this is ineffective parasitism—the glioblastoma inevitably kills its host—no more can be expected from such cancer cells; their programmed tactlessness and impropriety do them in.

The location of a brain tumor partly determines a patient's clinical course. Johnny's tumor flourished in relatively inactive occipital-parietal areas of his right brain. Apart from some loss of motor function on his left side and restrictions of his visual fields, he remained remarkably functional to the end. The tumor ultimately eroded a blood vessel wall, and the resulting brain hemorrhage killed him.

How very different was the clinical behavior of George Gershwin's glioblastoma: In February 1937, while playing his Concerto in F at Philharmonic Auditorium in Los Angeles, he suffered a momentary loss of consciousness. He missed a few bars, then continued as if nothing had happened. He later spoke of smelling burnt rubber. When physical examinations found nothing wrong, the incident was attributed to fatigue, the stress of his enormous success. In April of that year, in a barbershop chair, his momentary loss of consciousness and subsequent smelling of burnt rubber recurred. By June, he was suffering agonizing headaches and had become periodically listless, irritable, and confused; there were lapses of coordination; the smell of burnt rubber now haunted him. In this era that preceded neural imaging, his neurologist found no evidence of an organic lesion, and his signs and symptoms were attributed to hysteria. The glioblastoma had avoided clinical detection, living symbiotically with its host brain. It had probably been growing in this furtive way long before Gershwin's first loss of consciousness.

On July 9, 1937, Gershwin had a seizure and fell into a coma. Subsequent surgery disclosed a large cystic mass in the right temporal lobe; it involved too many vital brain structures to be removed. A biopsy revealed glioblastoma. He died several hours after surgery without ever regaining consciousness.

Despite the perilous ingress of his tumor, Gershwin composed two of his most beautiful songs, "Love Walked In" and "Love Is Here to Stay," in the last few months of his life. The processing of music is not as lateralized in adult males as is speech. Notwithstanding the volatile simmerings of Gershwin's right-sided tumor, his left brain could have assumed, over time, essential functions of his musical genius, allowing for his terminal inventiveness.

I see Gershwin, his neurons moving like piano keys, playing his concerto; his tumor cells press atop the neurons like so many thumbs, until the music stops.

B.R. IS A FORTY-FIVE-YEAR-OLD Caucasian woman who presents with a four-month history of headaches. A brain scan reveals a well-circumscribed mass lying beneath the skull and compressing her right frontal lobe. The neurosurgeon is able to scoop out this discrete bulk from the brain tissue it compresses. I receive it in three gray-pink, rubbery fragments that fit together as an oval mass about the size of an egg. I note the gentle protuberances of its surface, the lobulations of its cut edge. These are the overt features of meningioma—a benign tumor arising from the fibrous vestments of the brain. The microscope reveals a whorled growth pattern of meningeal cells that are as furled as a spiral galaxy. The nuclei are agreeably uniform, oval and blue, and the cytoplasm is faded pink, poorly defined. Many of these nuclei are so crammed with their own cytoplasm that they seemed eclipsed by it. An occasional amethyst calcification, concentrically laminated—a psammoma body—is visible, and clusters of xanthoma cells stuffed with lipid are scattered about.

Each tumor has its own life story—the nature of its cells, the imposition of its growth on surrounding body tissues, the threat it poses as an illness. This type of meningioma has a slow centrifugal growth that usually yields a globular or oval tumor. Reluctant to invade the brain, it displaces it. When favorably situated, it can grow insidiously beneath the skull to the size of a large lemon before producing symptoms. It is usually cured by surgical removal.

B.R. can expect a happy ending.

The benign swirls of this meningioma energize me. Although I am rarely part of a patient's emotional experience, I am not completely extricated from it either. The sight of malignant cellular disarray burdens me with all that it forebodes, gnaws at my own mortality. The vision of a benign tumor's orderly cell growth absolves me, makes me feel as if I myself have been granted a reprieve.

THE BRAIN CAN deceive those who would know it. I remember a thirteen-year-old boy who presented with seizures. Scans revealed a space-occupying lesion in his brain that suggested a malignant neoplasm. A needle biopsy was done under radiologic guidance. I distinctly remember the tissue I received; it was as white and friable as feta cheese.

Under the microscope, a spread of tubercle bacilli, the organisms that cause tuberculosis, appeared as minuscule blood-red leeches. Body cells in their wake, as if sucked dry of lifeblood, had disintegrated into amorphous fields of debris that stretched to the ends of the cellscape. Against a brilliant green background stain, these organisms resembled red tinsel heralding a high-colored microscopic Christmas.

In reality, the causative bacilli of tuberculosis act more like passive bystanders than active bloodsuckers in the wasted tissue of disease. With infection, the body's complex defense mechanisms slowly kick in to destroy these bacilli and the human tissues harboring them. In an effort to kill off tubercle bacilli, the human body is perfectly capable of destroying itself; it has done so for centuries.

I saw very few tubercle bacilli in tissue biopsies in the sixties and seventies. Despite its prevalence in Third World countries, tuberculosis was becoming an affliction of the past in the United States. It was alarming and disheartening, then, to witness a resurgence in the mid-eighties that remains with us today. And deadly new strains—superbugs—have emerged that are unscathed by all known curative drugs, inextricable as a spreading cancer.

Tuberculosis that masses in the brain—a tuberculoma—remains uncommon in this country. This biopsy surprised me; I had expected a malignancy. It was one of those humbling experiences that turns out well. The boy was started on antituberculous drug therapy, and within months the brain mass had disappeared. The biopsy *was* his microscopic Christmas.

Sometime later, I met the boy's father at the local library. He expressed his thanks to me for having been the bearer of such good news. He too had been expecting the worst and spoke of a feeling that God had lent a hand. He likened his son's diagnosis to a benediction.

TWO YEARS AGO I read a troubled brain in a whole new way. Although Parkinson's disease is not diagnosed by brain biopsy—the affected neurons are too deep, the structures too vital to warrant entry with a knife or needle—its essence hovers around me. I stare at the sepia photograph of my paternal grandfather, his facial expression enigmatic, his fingers pressed together in rigid extension like a clue. He and his daughter, my aunt Bess, both died of Parkinson's in their eighties.

After Bess's death, I went home to Montreal for my father's eightieth birthday. It was an exuberant celebration for an amiable man whose only concession to age was a rise in his golf handicap from single to double digits. A month later, an orthopedist examining my father's sore back thought he detected the staring visage of Parkinson's. His suspicion was soon confirmed.

The diagnosis of Parkinson's can be made clinically and supported with a PET brain scan. It results from a mysterious degeneration of specific neurons whose melanin-pigmented bodies lie in the substantia nigra, which, no bigger than a thumb tip, is a repository for these melanin-laden neurons. The neurotransmitter dopamine, a product of these dying nigral neurons, is also lacking. In this way, the computations of body movements are gradually jeopardized until the body is no more than a casement for the mind and soul. And no two Parkinsonians suffer their motor losses in the same way or at the same rate. Initially, Parkinson's body quirks can be subtle. Eventually, one or more cardinal signs appear to give away the body's morbid secret.

Rigidity is one cardinal sign of Parkinson's; it involves all voluntary muscles. And the poverty of my father's facial movements was an expression of that rigidity. The easy grin I had so often looked to for approval in childhood had permanently vanished as his stare hardened to a mask. It was as though a part of his personality had been stolen from him.

He was treated with Sinemet, which is converted in the brain to dopamine. Stemming his dopamine loss in this way seemed to halt the progression of his stiffness for a few years. At eighty-three, he stopped playing golf with his cronies and expressed a desire to play golf with my mother instead. "I'm making too many bad shots," he told her. "I tell my body to do something and it just won't do it." His hands now seemed too stiff to grip his golf clubs properly, and his postural reflexes were unequal to his task. Over time the Sinemet lost its effectiveness, so his doctor added Deprenyl to inhibit further dopamine breakdown. But drugs could no longer halt my father's increasing stiffness or the slackening of spontaneous movements. He was having trouble walking, even with a cane, and his dawdling gestures lacked dexterity. No longer able to hold cards properly or shuffle a deck, he declined bridge games with his friends.

Two years later, he needed a walker to visit his doctors. Otherwise, he was confined to home. He moved about with slow deliberateness, never laziness, and his daily routine—dressing, eating, using the bathroom, moving about the house, undressing—consumed the greater part of his day. His simplest rituals had become his life. A year later, his debilitation was such that my mother had to cut up his food and turn him in bed. Did he feel dehumanized by his body's dissolution, hopelessly bereft of his dignity? As a former athlete, did he view his rigidity as a bitter paradox? Did his burden dampen his sensuous pleasures, pepper him with despair? He never confided his feelings to me and regretfully, I never pursued them. Through it all, he remained mentally acute and uncomplaining. This stoicism seemed to attest to his courage and embody all that he endured.

In his eighty-seventh year, my father was hospitalized for a bladder infection. A sluggish functioning of his bladder muscles due to Parkinson's was no doubt to blame. He was joking with the nurses the evening before he was to be discharged, making sure that levity prevailed over illness. Later that night, while he slept, a massive heart attack killed him.

When I look back, my memories of my father are filtered through my understanding of the changes in his brain. I clearly recall the summer afternoons at his swim club. I was nine or ten years old. His substantia nigra teemed with healthy neurons then; their tapered bodies were so full of gold-brown melanin pigment that they resembled schools of trout. We'd buck the Saint Lawrence River's currents together, stroke for stroke. I can still feel that cold, choppy water numbing me, smell the riparian elms and maples and oaks, taste the sweet machismo of it.

I remember a day—I couldn't have been more than twelve—when I caddied for my father. The way he drove the golf ball off the tee is what stays with me:

There was a confidence in his compact frame, in his bearing, in the way he addressed the ball. His strapping nigral neurons were perfectly attuned, flawlessly promoting his athleticism. His backswing was shortened somewhat, and his stroke seemed easy. It was the hand speed he could muster that was extraordinary. The ball would explode off the tee, splitting the fairway, carrying below the height of trees. Rising slowly, steadily, borne in its orbit like a distant comet, it seemed to hang endlessly in the clear and piercing sky. And when it finally began its triumphant descent, it faded gloriously in gravity's hold. My father would look away in silence after these towering drives, pondering his iron shot to the green. He was all concentration as he charged down the fairway. I'd slide the strap of his golf bag across my shoulder pridefully and step up my pace alongside him.

Later, when television finally came to Canada, I remember the Friday-night fights. My father had been an amateur boxer in his youth and would perch on the edge of his chair, fists held taut in front of him. In a running commentary, he would instruct both fighters, infuse them with strategy. As the bout wore on, the winner always seemed to be listening to my father, while the loser seemed to pay him no need. Although he was usually alone in the room, his heated admonitions could be heard throughout the house. You knew if there was a knockout by the pitch of his fervent cries, and if you sat beside him, as I sometimes did, you'd witness his fists whipping through the winning combinations of lefts and rights. This ritual was my earliest introduction to instant replay and round-by-round fight analysis. Sometimes, if a championship fight went the distance, all that shadowboxing could wear my father down; but I think it was the rejuvenation that he yearned for. He still had his full complement of nigral neurons back then; they were pumping out their dopamine like geysers.

I spent my teenage summers as a stock boy in my father's factory, tallying fabric yardage in storage bins. It was nothing fancy: I would unroll the bolts of cloth and measure their lengths with a yardstick as dust clung to my sweaty skin. But mostly I watched my father's fierce struggle in this highly competitive arena, how it slowly bled the stamina from him. He went to great lengths to fashion dresses that had a certain style and flair, and somehow, morally bankrupt rivals were always prepared to copy his styles and undercut his prices. I'd watch him try to charm the buyers in his showroom, but quality and style were a hard sell in those days. Although he almost always maintained an outward calm, I know now that the hot-wiring of his anger circuits likely set his limbic system aglow. It was almost as if his nigral neurons got singed in the heat of all that stymied him. My father wanted me to succeed him in this business, but I knew early on that I didn't have the gumption for it.

My annual visits to Montreal during the seven years of my father's illness enabled me to see his slow freeze. Although his body ultimately imprisoned him, it highlighted the quickness of his mind. Often he seemed to use his rigid confines as a springboard, trying to reinvent and reclaim himself. It was this palpable spirit that made his helplessness so heroic to me. I envisioned his nigral neurons, now strangely toxic and fading, pocked with round pink bodies that looked like dabs of cirrus clouds at sunrise. As these neurons slowly died, incontinent of their melanin, I could see gold-brown granules clustering at neuron death sites, naked and alone as gravestones. And glial cells, ever at hand, blanketing the spaces left by the dead.

THESE DAYS I THINK about my own substantia nigra and those of my family. Susceptibility genes for Parkinson's occasionally have been documented in large families, though none so far in Ashkenazi Jewish ones like mine. Since no one in my family with Parkinson's is alive and can be tested for mutations, I have banked a sample of my blood. As new Parkinson's susceptibility genes are discovered, my sample will be checked for each newfound mutation. Do I want to know if Parkinson's lies ahead for me? Do I want to involve other family members, asking them to bank samples of their own blood? I see no harm in confronting head-on the threat of this impoverishing illness, and I hope to use the family discourse about it as a means to clarify what binds us.

My family tells me how I remind them of my father. Not only the physical similarity, they say, but the voice, the mannerisms. They smile when they tell me these things, and I wonder what memories of my father I trigger in each of them. Is this likeness intensified by my father's absence? Is it my own aging that reminds them more and more of how my father appeared in his healthy old age before Parkinson's became a part of him?

I superimpose my father's decimated substantia nigra on my own and wonder about the similarities. I have no pride, only dread, in any likeness here. In reverie, my face becomes thickset with age, and the grizzled hairs that sprout from my scalp and brows and mustache have me looking faded, exposed to time. Then my features freeze, a fixed stare that hardens against emotion. Behind this stony visage perks my lucid brain, its pulsing circuits orchestrating my angst and abject loneliness, my degradation.

If my nigral neurons do falter and Parkinson's entombs me, will other neurons, other circuits, brace me against my own looming rigidity? Will my father's courage and equanimity inspire me to move beyond my adversity?

For now, my nigral neurons seem pigment- and dopamine-loaded. Ticking with caution, I feel the silken surge of my body movements, sleek and nimble, flowing free. And my smile affirms all that moves me.

CHAPTER FOUR

Heart Rhythms

MAN IS THE SUM OF HIS RHYTHMS. HE reflects them in his bearing, movements, speech, emotions, even in his silences. They emanate complexly from his brain and the deep-chested beat of his heart. The quintessence of rhythmic man is the music conductor. One can hear and feel such a man's rhythm in his music; it is as if his other rhythms synthesize it. See the sharp sweeps of his arms, his expressive hands, as he exhorts the fourth-movement climax of Beethoven's Ninth or the soft deliberations of these same arms, hands, throughout the adagietto from Mahler's Fifth. And if his conducting, his music, surges entirely from memory, then Mehli Mehta is one such man. A colleague has introduced me to this nonagenarian and I visit him in his Westwood, California, home to learn how heart disease has affected his rhythm.

His contemporary house is of white-painted wood. The white picket fence out front opens to a winding brick path lined profusely with rosebushes, yellow hibiscus, scarlet bougainvillea, and touches of impatiens, begonias, camellias, nasturtiums, and chrysanthemums. Baskets of multicolored fuchsia hang from the porch roof like a textured tapestry. Mehli will later tell me that the sweeping gardens that engulf his home in so many vibrant colors are the music of his wife, Tehmina, her compositions, her symphonies.

"My Parsee ancestors were not Indian at all but Persians, followers of the prophet Zoroaster, believers in the god Ahura Mazda, Lord of Light," Mehli says. "Driven from their homeland around the seventh century by invading Arab hordes, they fled to the western shore of India, the nearest civilized coast. They settled in Bombay, mostly, where I was educated in my family tradition as an accountant. I took a position with the government as a tax collector."

We sit in his spacious living room filled with the bric-a-brac and mementos of a marathon, illustrious musical career. Photos and paintings of a seemingly happy family life are there. Mehli talks rapidly, an unrelenting staccato; this intensity of discourse is unusual for a man of his advanced age.

"Growing up in Bombay of the 1920s and early 1930s, I appreciated the melodies and rhythms of Indian music, but it was the harmonies of Western classical music that I heard on the wireless and the ten-inch, single-sided His Master's Voice records on gramophones, and in the salons of the British hotels, that really won my heart."

He is restless in his chair; it is as though it can barely contain him. And I am soon concerned that the unbridled vigor of his conversation is too much of an exertion.

"When I was seven I began violin lessons with local teachers," Mehli says. "And by seventeen I was studying under Oddonne Savini, a reputable Florentine violinist and local salon orchestra leader. Mozart, Beethoven, Brahms, Debussy. Puccini was just getting started then, you see."

He is leading me through his musical development rather than amplifying his thoughts on rhythm. Perhaps the two are intimately entwined.

"It was a passion I carried for my music that made me question my career in finance," Mehli says. "Then, in 1927, God sent Jascha Heifetz to Bombay."

His smile spreads now and his eyes are filled with a musical light. "He was on a world tour. I listened from the balcony; it was all I could afford. As he played his concertos and sonatas, *he* became God. What I had only heard on the wireless or phonograph, I could now hear and see in the flesh. It opened up for me a whole new world of possibilities." He pauses as if to let settle the flourish of his words. "On that afternoon in 1927, for a few precious hours, I *was* Jascha Heifetz."

I laugh, disarmed by his enthusiasm.

"The next year Efrem Zimbalist came. He gave six recitals and was every bit as godly as Heifetz. So in 1928, for another few precious hours, I *was* Efrem Zimbalist.

"Then came Anna Pavlova," Mehli says. "In 1929, she arrived with her international ballet company, conductor, pianist, violinist, and cellist. Hoping to recruit enough Bombay musicians for an orchestra and finding no readily available professionals, the company turned to me and some of my amateur friends." Mehli clasps his palms and fingers together. "Ahura Mazda had sent an answer to my prayers. I immediately left my tax collector post and established Bombay's first ballet orchestra. I became its concertmaster and principal soloist."

I begin to look around me at the casual piles of books and seventy-eight-rpm record jackets that fill the room. On the wall is a prominent oil painting of son Zubin frowning as earnestly as Beethoven, and a photo of Mehli's favorite composer—Brahms—which Zubin had copied from the original that hangs in Vienna Philharmonic Hall. Smiling family photos of senior maestro, wife Tehmina, sons Zubin and Zarin, their wives and children are there. Inklings of pride and unmistakable confidence.

"Pavlova came and went, but the orchestra remained to form the Bombay Symphony," Mehli says. "Had I been Hindu or Muslim, I would have banished the thought of a music career. The caste of musician in India was hardly higher than the pariah. But Parsees had no caste system, you see, so I became a professional violinist with all the blessings of my family."

Western classical music was a hard sell in India, and after ten years of trying to rouse more than a meager following in Bombay, Mehli came to New York in 1945. On a grant from the prominent Parsee Tata family, he spent his nights at Carnegie Hall in thrall to the violin virtuosos—Menuhin, Heifetz, Zimbalist, Milstein, Kreisler. In time, India's best violinist came to a startling realization:

"'Mehli,' I said to myself, 'God has not given you the fingers of a virtuoso, but he can certainly help you hold a stick.' So I took myself to Carl Fischer's music store and bought a two-dollar stick and dozens of study scores. I returned to Carnegie Hall to study the styles and conceptions of Koussevitzky, Toscanini, Szell, Reiner, Ormandy, Stokowski, Monteux, Walter. They were all there, the world's great conductors. My New York friends would go to the concerts for enjoyment, but for me they were master classes. Here is where I learned to be a conductor."

I am sensing how Mehli's rhythms begin with his heart—so much of his heart—and fuse with the rhythms of his music; they are his vitality, his tenacity, his virtuosity; they are all of his possibilities.

THE HUMAN HEART glimpsed through a cleaved rib cage in the course of surgery, obscured by its membranous mantle, thrusts with an urgent insistence. It is not until the mantle is opened that you clearly see the muscular glory that is heart, how it beats with the smooth imperative of a thoroughbred. What keeps this organ contracting through chaos and calm for so many years without a falter? Surely it is remarkable for the marathon distance of its run, propelling the blood as it does through myriad arterial arbores.

And what of its beat, the fractured sound the heart makes deep within the darkness of its crib? Lup-*dup* . . . lup-*dup* . . . lup-*dup*. Iambic patter; the second sound is stronger, as if to underscore the first. It is the pulsing of a drum marking our time. The body's metronome. Unlike the muscles of our face or trunk or limbs, which move to the stimulations of extrinsic body nerves, the heart contracts (and beats) of its own volition. Wear it on your sleeve and see its autonomous thump. The lup and the *dup* and the resonant silences that fill in between are the audible culminations of the complex movements of all its integral parts. To know a heart's sound, like that of a car's engine, is to know that heart, that car, itself.

And what are these integral parts? Cylinder-shaped heart muscle cells (myocytes), brimming with contractile proteins, compose four heart chambers:

The upper two atria—smallish oranges—sit atop the lower, larger ventricles that taper like a stemless upside-down artichoke to an apex, completing the valentine. Four valves connect these chambers, guiding the blood along its abiding path, preventing a backflow. And the beat is triggered by collections of electrically excitable myocytes that rapidly spread synchronized waves of current across the heart. In the course of this electrification, myocytes pull in unison with their neighbors until the chamber muscle they compose contracts, launching the blood like a massive bleed.

Put your ear to a loved one's chest and listen to the heart. It is the closing of valves that makes the heart speak; they are its vocal cords. And if all is not well, these valves can groan, or wail, or helplessly sob. But so long as we live, they are never silenced.

When fear prevails, the heart does not climb into the throat—it cannot leave its vesseled bed—but throbs instead with clamant fury. And when the body cools, the heart slows as if hibernating in its dark cave. It is the brain and its monumental tails of nerve that tweak the pace of the heart, slow it down or speed it up: *ritardando, accelerando.* But in matters of rhythm, the heart has its own mind.

IT IS THE HEART that captures our imagination and not the myocyte. And yet the heart is filled with cellular drama: Cylindrically shaped myocyte cytoplasm is chock-full of long protein fibrils that contract the cell; a liberal scatter of mitochondria are there to supply the high-energy demands of the unrelenting fibrils; and one or two cigar-shaped nuclei, repositories of genetic blueprints that govern the cell's function and form, are centrally placed, prominent icons. And enclosing it all is a scalloped membrane and terminal disks that allow for the electrical flow that launches the contractions. It is the myocytes that build the hot-wired heart, each one a member of this unique electrical symphony. And the potential disaster of an electrical fallout of these cellular musicians from coronary heart disease has been sorely minimized.

Premature beats, short circuits, chaotic electrical cadences alter a heart's regular rhythm. These are arrhythmias, myocyte electrical malfunctions. A modified rhythm can be harmless or lethal, simple or inordinately complex. As many as five hundred thousand Americans die each year of lethal arrhythmias, a disease that claims more men in the Western world than any other. Arrhythmia victims die too suddenly to make public their catastrophes. And this heartrending illness endures, largely ignored.

AT AGE SIXTY-ONE Mehli Mehta had his first heart attack. His progressive coronary artery disease (arteriosclerosis) had resulted in a sudden, focal coronary artery blockage, and the subsequent blood and oxygen deprivation to a portion of his heart destroyed the myocytes there. They were replaced by the fibroblast cells of scar. "I was told to go home and be quiet. But can I be quiet? Of course not," he says. His grin has the mischievous assurance that successful survival for ninety-one years allows. "I was so obsessed with conducting then that I was leading five different Los Angeles orchestras. So two months later I did five concerts."

Mehli has conducted major orchestras the world over, debuting to critical acclaim at Carnegie Hall with the National Orchestra Association Symphony. At the helm of UCLA's orchestra department in 1964, he founded the American Youth Symphony; he has been its principal conductor for thirty-three years. It is this orchestra that will be his legacy.

"I should have rested, because a month later I had another heart attack. So then I listened to my doctors, restricted my diet and observed caution in all things. I was fine for almost thirty years, until 1995."

Behind all the technical mastery was a desire, if obsessive, to express his strong affirmation of life through classical music. He had to perceive his illness as a *major* threat if it was to make him change his lifestyle.

It was in 1995 that Mehli had his third heart attack; it culminated in coronary artery bypass grafting. The arteriosclerotic, scarred heart has a flair for the dramatic: It is prone to life-threatening arrhythmias, and Mehli's battle-marred pump was no exception. The following year he endured a sudden bout of unnerving palpitations. He was rushed to the medical center at UCLA, where a life-threatening arrhythmia was observed. At the edges of Mehli's scars, where the rigid splay of fibroblasts entwine the viable myocytes, electrical waves had run askew, and random short circuits dominated the rhythm. Rapid beating was preventing his heart chambers from adequately filling with blood; its thrusts were too incompetent to promote the circulation. A man can quickly die in the throes of such heart frenzy. Unless this frantic rhythm reverts spontaneously to normal—and Mehli's did—the heart must be electrified from without, reminded how to use its forsaken electrical conduction paths to rekindle its potent rhythm.

As a precaution, an AICD (automatic implantable cardioverter defibrillator) was implanted beneath the pectoral muscle in Mehli's chest and wired to his heart. This new-millennium, programmable microchip device is sophisticated enough to detect and arrest life-threatening arrhythmias. Should another occur, the AICD is designed to deliver an electric shock—several if necessary—to Mehli's heart, jolting it back to life.

It is a sad fact that cardiologists are as yet unable to predict the onset of many arrhythmias; they get to treat only those hearts that survive their initial assault.

WHERE IN THE HEART lies courage, or love, or grit? Deep within the thickness of its walls, tucked beneath the cusps of its valves? How the heart can obfuscate!

Mehli Mehta hands me a videotape, his farewell story. It is Saturday night, May 16, 1998. Royce Hall, the iconic UCLA auditorium, empty for four and a half years in the course of earthquake-damage repairs, is reopening its doors. The triumphant return of the American Youth Symphony (AYS) orchestra is tinged with sadness; its indomitable maestro is to conduct his farewell concert. His battered heart is no longer equal to all its rhythmic tasks.

The magnificent Skinner organ, an integral part of Royce Hall since 1932, had been severely damaged. Mehli has decided to showcase it anew with Saint-Saëns' *Organ* Symphony. He further shapes his farewell concert with a pair of demanding display pieces that these AYS musicians have never performed, Ravel's Second Suite from the *Daphnis et Chloe* ballet and "La Valse."

Mehli walks slowly, deliberately, to the podium, bowing appreciatively to resounding applause. He takes his position as 106 young musicians fix their gazes uniformly upon him. He has positioned a bar stool carefully behind him should he tire.

An expectant hush . . . and we are into *Daphnis et Chloe.* I listen to the clarinets and flutes bubble the earliest dawn. . . . This videotape sound does not do justice to the impeccable Royce Hall acoustics. Now a vibrant wash of strings calls forth the earliest rays of light. The tessitura in which the strings are playing gets higher and broader as dawn is clarified into morning sunlight. The maestro's movements are clear, precise. The right hand to rhythm, the left to expression, evocation, nuance. Soon he lifts his left arm triumphantly to signify the crescendo of a full sunrise . . . and the melody becomes as languid as movement in summer heat. The scene flicks—an edit—and abridges; Mehli continues in vigorous form: concentration, precision, stamina, polish . . . the final flourish.

A handshake for the concertmaster. A gracious bow to fervent cheers. A confident grin. A springy amble off the stage. Maestro returns forthwith for "La Valse." This piece is complex. Ravel has distilled some of the waltz music conventions here, reshaped them. What remains is scarcely a waltz. Mehli is conducting in three; he moves his right arm, his hand, to each beat. Is he working too hard? Perhaps the orchestra is slowing him down so that he cannot conduct enough of this piece in one—one arm, hand motion for every third beat. So much conducting in three seems a test of his stamina, a drain of his energy. Does he do so to keep his fledglings together, to carefully manage the tempo? I do not know. Clearly this is a strenuous composition to play and to conduct. "La Valse" ends with a sustained flourish. Maestro's arms are in full swing like an aerobics instructor's. But what of his heart?

A handshake for the concertmaster. A bow to fervent cheers. Is that a grin or a grimace? His gait seems a bit unsteady now as he slowly leaves the stage. But he does not enter the wings. He turns about hesitantly and returns to the podium to heartfelt applause. This is unusual: A maestro characteristically leaves the stage after a performance and, out of sight, is wheedled back by persevering applause. Mehli has cut short his appreciation scenario.

What is about to happen is startling. A bouquet of flowers rises from the bottom of the screen, clutched in the hand of an ardent admirer. Reapproaching the podium, Mehli leans forward to receive these flowers and is suddenly jarred from within. His knees almost buckle. He grabs the podium with his left hand and clutches his heart with his right, his baton still propped in his fingers. The AICD has just discharged, jolting his heart with electricity. For the briefest moment the world comes to a stop. Mehli's arrhythmia seems to revert to normal as quickly as it began. He smiles to unceasing applause, bends to receive his flowers, places them beside the stool.

A few seconds pass and the AICD strikes again. This time the baton is in his left hand and he punches his chest with his right fist, in anger or mortal fear. Mehli's cry is guttural. As if in response to this mournful plea, his heart presumably returns to a normal rhythm and pace, and maestro walks uneasily off the stage.

Intermission.

MY MEETING WITH Mehli is long after his farewell concert. In the quiet of his house, he reflects on what happened that fateful farewell-concert night, how his heart rhythm had twice gone awry during the intermission. He had narrowly avoided a melodramatic fate. The details of his experience were supplemented by his cardiologist and by examination of his AICD. I sense that Mehli has told this story many times. His manner is superbly unburdened; only his eyes reflect his secret, adamant strength.

"It was as if I had been kicked twice in the chest by a horse. I can't describe it in any other way," Mehli says. "After that I went back to my own life."

Not quite.

He was very lightheaded as he walked from the

intermission stage, he tells me, and slumped into a chair in the wings. Zubin, having flown in from Florence for this momentous occasion to surprise his father, ran to his side. With him was a cardiologist friend of the family. They supported Mehli by his elbows, walked him to the green room. There they removed his coat, loosened his tie and collar, and laid him on his back on a couch.

"I was half dazed and very weak but so delighted to see Zubin. I had no idea he would come all the way from Florence," Mehli tells me.

I try to envision the green room. Tehmina was there, and Zarin, and Zubin's wife. The cardiologist did not allow others into the room. He monitored his patient by repeatedly taking his pulse.

"They were all afraid that I was going to die there, but I knew better. Zubin had come to me like a vision. I knew God had sent him to help me back to life."

In fact, the cardiologist knew that the AICD had done its work. Twice shocked, Mehli's heart had reverted to a normal rhythm and a pulse rate of ninety beats per minute. But what to do? Mehli was beginning to feel better, to talk about going back out for the Saint-Saëns.

"Daddy, you can't go out there now," Zubin said. "The Saint-Saëns is forty-five minutes long. It's too much for you now. We're going to make an announcement that it has been canceled."

"That is just like my son," Mehli says to me. "Whenever he is in the room, no matter where, he is in charge. Everybody follows.

" 'No, no, darling,' I said to him. 'My orchestra is onstage. The audience is waiting. Dr. Harmon, the organist, had to rent an organ because the Skinner is not yet ready. It cost him a thousand dollars.'

" 'No, no, Daddy. It is not possible,' Zubin said.

"Then the cardiologist spoke words of wisdom. Just like Solomon, he told me that I may conduct the Saint-Saëns, but only if I remained seated.

"I didn't mind sitting so long as I could do my job. If I was going to die, conducting the Saint-Saëns *Organ* Symphony was not a bad way to go. So after thirty minutes, I brushed my hair, straightened my tie, put on my coat, and went back out there. And I told Zubin to go back to his seat. He was the one I was conducting for anyway."

Throughout the Saint-Saëns, unbeknownst to those in the audience, the cardiologist and a pair of paramedics waited anxiously in the wings. The remainder of the concert went off without a hitch. Said Daniel Cariaga of the Saint-Saëns in his *Los Angeles Times* review, "Without raucousness or acoustic overload, the 106-member orchestra and its tasteful organ soloist filled the renovated hall with beautiful sounds and controlled resonance. Mehta brought freshness and élan to a very familiar work; his younger colleagues gave him a full palette of colors."

Upon interrogation in the cardiologist's office, the AICD revealed that Mehli had experienced a sudden, rapid heartbeat. Timed at 180 beats per minute, his runaway heart had forced the maestro to cut short his appreciation scenario. His heart frenzy was then terminated with a thirty-one-joule jolt. When it recurred after two heartbeats, a second shock of equal magnitude terminated it permanently.

Mehli has not had a life-threatening arrhythmia since his farewell concert, and if he should, he is completely confident. He carries his own mechanical paramedic tucked neatly inside him.

TWO YEARS HAVE PASSED since Mehli Mehta's fateful farewell concert. Mstislav Rostropovich has been booked to play with the American Youth Symphony in support of cardiac research at a local medical center. Mehli is to conduct this all-Tchaikovsky program. But there is a problem. His scarred and failing heart is bolstered by medications. The stress of a performance, no matter how felicitous his skills, could precipitate a lethal arrhythmia and kill him. His wife and sons are adamant: No more conducting. The time for discretion is at hand.

Mehli has a different view, a different set of values. "Spencer, dear," he says to me, "an artist can never stop; he must create more things, go places musically where he has not been. I have played with all the great soloists of our time except Rostropovich. I said to him, 'Slava, you must play with me before I die. Don't stretch me too far. Come soon.' "

Rostropovich finally comes to play with Mehli Mehta and his AYS. And so the larger drama unfolds.

"Only for Slava would I risk my life to play," Mehli says. "If he did not come, I would have listened to my family. It has nothing to do with hospital fundraising; if Slava comes to play with me, I'd play in the jungle."

It is a lucky man who so loves what he does. Instinctively he gathers in his courage, commitment, talent, to feed his insatiable need to make classical music. For Mehli, to conduct the AYS in Variations on Rococo Theme, op. 33 for violoncello and orchestra, with Rostropovich as its soloist is a musical risk, a need for harmonic exaltation that takes priority over hazards to bodily organs.

Two weeks before the concert, Mehli's failing heart becomes arrhythmic. Short of breath and fatigued, he is admitted to the hospital where his heart is electrically converted to a normal sinus rhythm. He remains in the hospital for a few days where, propped in his bed, eyes closed, he is moving his hands, conducting the Tchaikovsky concert music that teems in his head. "After this concert, I don't care what happens to me," he tells his cardiologist. "But keep me going until it's over."

The cardiologist, awed by this nonagenarian's courage and sensing how entwined Mehli's music is with his will to live, cautiously supports him in his latest endeavor.

Concert day. Mehli's heart remains in a normal sinus rhythm. I am invited to the morning rehearsal at the Beverly Hilton Hotel. Some of Mehli's friends and family are there. From up close, I see Mehli sitting on his stool, conducting an auspicious gathering of pubescent musicians. They begin with the Andante

Cantabile for full strings, and Mehli interrupts them from time to time to subtly tweak their sound. To me, as I listen entranced, wrapped in their music, they have the balanced sound, lyrical elegance, poetry, of a major symphony orchestra.

At the conclusion of the Andante Cantabile, Rostropovich appears onstage. He is a vigorous, athletic-looking man in his seventies. They begin the Variations on Rococo Theme, and as I watch and listen to Mehli conduct with his friend Slava sitting beside him—a master of his cello—I am swept by a tenderness for them both. Age has not tamped their music-making; perhaps it has enhanced it. The concert concludes spectacularly with the *Romeo and Juliet* Overture—Fantasy. There is a smattering of applause from those of us fortunate enough to be at this rehearsal, and we stand in line onstage to shake Mehli's hand. When it is my turn, I realize how frail he is. His hand is cold when I shake it. "Tonight will be the test," he says with a sense of urgency, and I am uneasy about his welfare.

"Of course it went well," Mehli tells me one week after the concert. "My oxygen was backstage, my doctors were there."

His cardiologist sat no more than fifteen yards from Mehli's conducting podium. And despite the fact that Mehli had his AICD tucked inside him, the cardiologist brought an external defibrillator to augment what he could do should an intractible arrhythmia stop the music. Mehli could not have conceived of a safer setting for his concert. Being a hospital fund-raiser, the event was packed with cardiologists.

"It was amazing how he mesmerized everybody in the audience, and how articulate he was in introducing himself, his music, and his celebrated soloist," his cardiologist tells me. "His extemporaneous remarks were measured yet extremely powerful."

I am relieved to see Mehli back in his house. The sparkle has returned to his eyes and his grin is full of mischief. He tells me that he will do yet another concert when he turns ninety-five. (He is almost ninety-two.) Just like Stokowski.

I ask him what he plans musically and he shrugs. Ahura Mazda will have to keep his heart in tow for another three years. There is plenty of quiet time for him to plan his next extravaganza.

CAMUS SAYS OF SISYPHUS that his infinite lifetime of futile labor is a meaningful one, that repetition is, in effect, the story of human life. Insofar as we live our lives in the moment, throw ourselves into our work, Camus assures us that we can, like Sisyphus, enjoy our lives. If we accept this philosophy, can there be any happier organ than our Sisyphean heart? It is precisely by its repetitive thrusts that it achieves validity. And what other organ can live so much in the moment when it swells with love or breaks from it? But let the brain come in, brimming as it is with reason, and it can bring a heart down, twist it until every vestige of electrical purpose is gone.

I listen to my own heart: Every now and then it speaks to me; it skips a beat. It has been palpitating in this way for as long as I can remember. In this hesitation to contract, I sometimes sense a falter. For a fraction of a second I am without heart. This pause is a chance, I believe, for my heart to take the measure of its heartscape, the cascade of its electricity, the mystic alliance of its myocytes, the skein of its emotions and passions, before moving on. Such palpitations in an otherwise healthy human are usually not a threat; in the large electrical and physiological scheme of things, they are inconsequential.

About a year ago my own cardiologist wrapped me in a twenty-four-hour Holter monitor to see what else my heart had to say. Wired, the pocketbook-size monitor strapped to my side, I was to carry on with my "normal" day.

I worked as best I could. I wrote and, fettered, finally fell asleep. At 4:40 A.M. the telephone woke me. My sister was calling from Montreal. Her husband, in the throes of chemotherapy for acute leukemia, had acquired an opportunistic fungal infection and quickly, unexpectedly, died. My brain's anger, shock, sorrow descended mercilessly upon my heart, and the monitor recorded it all. A few hours later, instinct got me to the YMCA. I ran on the treadmill, got my heart racing, my blood gushing like torrents of tears through untold vascular spaces.

I returned the monitor, noting on the questionnaire the grim nighttime telephone call, the early-morning workout.

The next day my cardiologist called. "Your monitor findings are a case report," she said.

"Me? What's so reportable?"

"Your palpitations are of little consequence, but you illustrate beautifully the heart's very different response to mental and physical stress."

Although every heart tracing tells a story, how could I suspect that mine would be so cardiologically compelling? When I was suddenly roused with terrible news, it was mental stress that involuntarily prodded my heart to a rapid 150 beats per minute. Ventricle myocytes must depolarize and repolarize—contract and relax—in unison for maximal pumping efficiency. With stress, repolarization rates of the ventricle myocytes can vary. Such variation is inefficient; it can set up short circuits even in the normal heart. The stressed brain simply cannot dabble with the devoted heart in this way; it can cripple its rhythm.

Then I exercised at the Y. As my heart rate slowly built to 144 beats per minute, its behavior was quite different, less mandated by the thinking brain. There was a smooth flow of thrusts, rapid relaxations in sync with rapid contractions. There is no electrical risk in the swift synchrony of a normal heart's physical stress: here the heart is potently engaged; it can run with abandon.

It is well known that mental stress increases the risk of short circuits, of lethal arrhythmias in people with documented heart disease and in those who are congenitally susceptible. What is not generally known is the lesser risk of sudden cardiac death from lethal arrhythmia in certain people with unidentified heart problems. Under mental stress, the normal-appearing heart in these people may be subjected to dangerous and complex electrical alterations.

If my mind's stress has my heart racing and pulse waves pound my body's shores, I close my eyes and deepen my breath. Transferring the rhythm of my heartbeat into a drumbeat, I let the rhythm of my music begin. An aria, a chanson, a simple lullaby, and soon my protean mind changes its rankling tune. Edginess dwindles and myocytes blend electrically together, once more in decisive harmony.

CHAPTER FIVE

Early Alzheimer's: A View from Within

I FIRST ENCOUNTERED ALZHEIMER'S DISease (AD) at the tender age of ten. While sleeping over at the home of my maternal grandparents, in their den, I was suddenly awakened by an apparition. Bolting from my bed, I came face-to-face with my grandmother. Her missing teeth—she soaked them in a glass on her bedside table while she slept—left her lips devoid of a supporting structure, and as she screamed at me, her mouth imploded.

"Tell me who your Zayde is running around with. I know he's seeing another woman."

As she stood by the large window that opened onto Sherbrooke Street, the chiffon curtains blew in around her like gossamer wings and I saw, in the faint reflected light of streetlamps, the madness in her eyes.

The fierce irony of her mistrust was not lost on me, for I knew even then how my grandfather doted on her.

"Zayde has no girlfriends," I said. "I promise."

My words did not comfort her, and as I moved to help her back to bed, she was out the door. So ephemeral was our nighttime encounter that I awoke in the morning convinced that it had been a bad dream. But the delusions and paranoia of my grandmother's illness were real. Alzheimer's disease was claiming her mind.

As an adult, I had always thought of AD in this way: a progressive mind loss of obscure origin with characteristic clinical and pathologic features. True, there are variations in the patterns and progressions of this illness, but they always seemed to me of little consequence. If AD mind loss had a beginning, an early stage, I believed it was but a short-lived step in a progressive mental incapacitation that jarred the victim from his life.

It was not until I met Morris Friedell that I came to see the flaw in my AD conceptions. He showed me how a regressing mind, like a deteriorating body, can crystalize the life that remains and devise ways to enhance it. I had taken the mind to be an all-or-none entity; I was unaware of the possible magnitude of man's mental capacity in the face of significant neuron losses.

"MY SHORT-TERM MEMORY HAS ALWAYS BEEN unreliable. I didn't find myself losing keys when I developed AD because ever since my childhood of losing keys, pens, and mittens, I've coped by using procedural memory," Morris says. He is a small and gentle man with a copious, frizzled beard, a sociologist retired from the University of California at Santa Barbara, who looks younger than his sixty years and has a professorial felicity with words. "When AD made me more forgetful, it was easy to expand my ingrained habit of relying on rules, and I shifted the burden to my procedural memory, with its alarm watch and notebook prostheses."

Having lived with his AD diagnosis since September of 1998, Morris was brought to my attention by a nurse practitioner friend who singled him out for his high-minded approach to his illness. Morris's description of his pre-AD vulnerabilities gives me pause.

Are there specific traits in some of us that are predictors of AD, traits that are discernible yet whose import goes unrecognized? Epidemiologist David Snowdon studies nuns. He and his colleagues have analyzed nun autobiographies written in their subjects' twenties—prior to taking their vows—for idea density (the number of ideas in every ten words of writing). They reevaluate the nuns in old age. Those whose autobiographies have the lowest idea densities are the most likely to develop AD. This study, still too small to be conclusive, is nearly 90 percent accurate in predicting AD. Despite the singularity of Snowdon's population, he strongly suspects that any youthful writer of simple sentences, ones that lack a complexity of interrelated ideas, is demonstrating a cognitive handicap predictive of AD.

The Nun Study is startling to me. I cannot help wondering about the idea density in the early correspondence of our fortieth president, Ronald Reagan. Can we find evidence in it that portended his AD? Are the small errors and inconsistencies in Iris Murdoch's last novel, *Jackson's Dilemma,* a harbinger of the AD that was becoming clinically apparent at the time of its publication? Is this novel muted in any way by a recognizable failure to simultaneously process a sweep of ideas? Can we compare the density of ideas in Picasso's paintings with those of his contemporary Willem de Kooning, whose later paintings, when his AD was acknowledged, might convey a certain "spareness"? Can we analyze our own documents written in youth and know that we will or will not develop AD? There would seem to be no end to the questions raised by the Nun Study; the reality is that its findings have yet to be confirmed with individuals of different lifestyles, habits, and risk factors.

"I'm much less concerned about rehabilitation for forgetfulness—I can take notes for such things—than about retaining my personality," Morris tells me. His eyes reflect an intelligence that clearly remains as he describes his evolving incapacity. And he appears to enjoy taking the mental measure of himself. "It seems that as long as a basic repertoire of emotions and an adequate level of self-awareness are retained, personality can be reconstructed," he says.

Personality—traits of character and behavior that constitute some measure of our mental processes—is a part of the brain's hardwiring, or ingrained circuitry.

Alterations of the personality's hardwiring are common in AD, often leaving a person disinterested in his surroundings, and his social behavior inappropriate. Morris understands the neuron losses, the brain circuit-processing difficulties that lie ahead as he strives to maintain his personality; he is aware, too, that little allowance is made for those who lose it.

We are lunching in an International House of Pancakes (IHOP) in Ventura, California, where Morris lives. He is moving to Livingston, Montana, in a few days, and I try to make the most of my brief visit with him. I ask him how he perceives his neuron loss, how he *feels* it.

"Thoughts no longer percolate my brain; they've slowed, become viscous," he says. "This has deprived me of the capacity to relish experiences, though I can still savor them."

I imagine thought "viscosity" as a blockage, but Morris assures me it is more analogous to mental fatigue, a lost buzz of interconnectedness. It is as if his mind is functioning underwater. Although his intellect remains largely intact, the rapid processing of myriad related thoughts that we all take for granted is slowed. "My thoughts occur in a linear, step-by-step fashion, as if my inner computer no longer opens to more than one window at a time," Morris says, as though sensing the contrasting rapidity of my own thoughts. "Dealing with multiple windows creates a disorienting circuit overload."

AS A SURGICAL PATHOLOGIST I am less attuned to matters of personality in Alzheimer's than to the mysterious AD aberrations of specific brain neurons. I look to the sweep of Morris's forehead as if to see beneath it the convolutions of his brain surface, how they have begun to dwindle and the furrows to splay. With my microscope, I see the typical exotic inconformities in AD brain biopsy neuron structure. They are as compelling as they are puzzling: Lashes of the neuron's nuclear eye appear coarsened, clumped, and tangled with what looks like cellular mascara; classic extracellular plaques, resembling bull's-eyes, can transform the cellscape into a simulated shooting range; artery walls can be thickened with homogeneous pink amyloid protein that appears, in cross section, as round and thick as a bagel. These curious imperfections are not specific to AD; they can be found in other dementias and in smaller concentrations in the normal aging brain. What defines AD at the microscopic level, without so much as a conclusive causal explanation, is the flourish of cellular tangles and extracellular plaques that violate the brain surface neocortex and the hippocampus below, associated always with assorted deposits of amyloid.

All that Morris does to infuse his fading cognition, his personality, lies beyond the microscopic alterations that I see; these are diagnostic markers that, as yet, shed little light on the exacting complications, the fears and bewilderments of AD mind loss.

MORRIS QUOTES FROM Oliver Sacks's book *An Anthropologist on Mars: Seven Paradoxical Tales.* In "The Last Hippie," an essay about a young man who had suffered severe brain injury from a tumor, Sacks writes that "he lacked the constant dialogue of past and present, of experience and meaning, which constitutes consciousness and inner life for the rest of us. He seemed to have no sense of 'next' and to lack that eager and anxious tension of anticipation, of intention, that normally drives us through life." This description seems to Morris to capture his own mental state. He rises each morning and prepares a list of his day's forgettable events; then he crosses off each one as he lives it. Without his docket, he can be oblivious of events that await him. His dialogue between past and present has been irrevocably altered, and I wonder, are his emotions braided with his other diminished faculties? Are they altered, too, in some way?

"My emotions seem less vibrant now, more detached somehow, yet a burst or a gush of them can sometimes part the cognitive clouds," he says. "If I'm ruffled by something I say or do, forgetting someone's name, for instance, the embarrassment I feel can stimulate me to think of something quite clever and appropriate to handle the situation."

Initial AD disability as mild as Morris's more often causes a person to feel anger or frustration or fear; paradoxically, this can lead to a functional decline and social isolation that far exceeds the disability itself. An emotional well-being thus seems paramount with early AD, and I wonder what role psychotherapy has to play.

"It's sad to me that supportive therapy is very much a part of the AD treatment profile, but deep-grief work and future-oriented therapy are not," Morris says. "Seemingly nothing can be done except to make us comfortable while we sit around and rot."

Unfortunately most patients with early AD are

not as insightful as Morris is, and they are often apathetic. For those with early AD who share Morris's acumen and verve, a future-oriented mind-set would seem an ideal tack for rehabilitation.

Morris's residual cognitive abilities—in particular the long-term memories spared so far by AD—allow him an extraordinary exercise.

"I pick two books at random from my collection, books that have long held meaning for me, and hold them in my mind, attempt to compare and contrast the ideas they convey," Morris says.

In the course of our conversation he refers to some of his books: He quotes Viktor Frankl on the meaning of suffering (*Man's Search for Meaning*) and Anatole Broyard on the need to develop a style for one's illness (*Intoxicated by My Illness*) and Marianne Paget on the molding of an invulnerable human spirit (*A Complex Sorrow*). Any comparison of such books, written by writers who have looked inward to find meaning and insights amid the harshness of their own outer worlds, would no doubt have galvanized his former sociological mind. Now this seems an exceptional way to explore a fading intellect. To the extent that the ideas in these books, the values, can orient Morris to the internal world of experience and supplant the external world of achievement, comparing them must be therapeutic.

We drive to the Alzheimer's Association offices near Morris's soon-to-be-vacated home, and I watch and listen as he visits with the staff. It is remarkable to hear him discuss, two years after his diagnosis, the nuances of his illness. If I did not know that his AD diagnosis had been made at UCLA, that his PET brain scan was corroborative, I would be hard-pressed in the short span of time we have been together to detect his organic brain impairment. In fact, he is remarkably focused and full of charm.

"I usually pass for normal, if a little absent-minded," he later tells me. "But I must pace myself. I tire easily and can mentally lose it."

The afternoon passes and I am increasingly aware that Morris is working with a practiced caution and concentration to achieve his demeanor. It appears to be exacting work, a drain of his energy, to pass for normal. We leave the Alzheimer's Association and I drive my car to the south ramp of the freeway. Morris quickly offers a course correction. I sheepishly turn north.

Folded sheets of paper appear and reappear from his pockets as he edits his notes and lists, deletes the events that have passed and prompts himself on what is yet to come. I wonder if he uses subtler prostheses to enhance his thinking. Apparently so. As he talks with the ladies at the Alzheimer's Association, I realize that Morris is not initiating conversations, he is "reacting" to what is said. He smiles when I question him about it.

"I usually communicate by computer. It doesn't matter if what I wrote five minutes ago is lost in the fog. I can simply scroll up to get it back," he says. "But when I have a conversation, I usually piggyback on the other person's train of thought. People don't notice then that I'm impaired. Instead, they're often flattered and can feel as though they have ingratiated themselves."

This is true. He fixes on my thoughts and expands them. He epitomizes the good listener, and I come away enhanced by the interaction, assured that he is stimulated, even gratified, by all that I choose to discuss with him. It is remarkable to me how our neurons can brace themselves on those of others and, supported in this way, enhance their own potential.

THERE ARE A FEW early-AD individuals who are active on the Internet, in the Memory Loss mailing list.

They form a support group. Here Morris writes an essay that he shares with the others. In it, he notes that "so many people go through life like tourists with a camera between their eyes and the world. With early AD, the linear processing of thoughts slows a person down so that he may learn to really see, to prize."

Laura replies, "Thanks Morris, I've noticed that I have a large amount of appreciation for whatever I'm focused on, it is very clear and real. Look away and it is gone. Look back and it is fresh and new. I am checking this out with a red geranium blossom right now. When I look away, 'red' no longer exists except as an abstract term. No blossom image remains . . . but I can look again."

Morris is enticed by his exchange with Laura and, aware that time is limited, decides to fly to Montana to see her.

"When Laura and I finally met and touched," Morris says, "I fell in love in a way I would never have imagined, as a person who has only known noon cannot conceive the pink, gray, and purple glory that sunset brings."

There seems to be nothing blunted about this emotion, but Morris sadly asserts that there is. Nevertheless, they are making plans to live together in Laura's Livingston home; she is working with building contractors, enlarging it to accommodate Morris and his stockpile of books.

This afternoon, after I leave, Morris is sending the boxes of his books on ahead and closing up his Ventura home. We drive from the IHOP to his friend's house and I bid him farewell.

I feel the oddest sense of loss. I have known Morris for only a few hours, but I will miss him. His sober resolve stirs the languid depths in me. His heroism, he tells me, is not in facing the fear of mind loss but in accepting the reality of it and adapting. He is actively participating, with all the resources he can muster, in the deadliest of mind games.

I will miss Morris for another reason: As he piggybacks on my train of thought, so do I on his. It is my attempt to learn as much from him as our brief time together allows. I feel a rare sense of intimacy with him, one that I am hard-pressed to achieve even in the closest of my relationships; it involves a careful, almost tenacious listening to one another, as if each of our worlds depended on it.

ALTHOUGH A PERSON with AD is much more than the sum of his tangles and plaques, these microscopic curios pose an interesting problem. Are they the neuronal glitches responsible for mind loss or are they the brain litter of yet unidentified pathologies?

Recent research suggests the tau proteins normally cementing neuron subcellular structures crumble in AD; the twisted tangles are the remains of the cellular dying. And what makes tau tangle? No one knows, but some suspect that the plaques are to blame, that these toxic protein fibril clusters sitting outside the neurons can "rub them the wrong way," setting off the internal destruction.

The neurotransmitters of AD-affected neurons are also found wanting. The loss of acetylcholine, for example, and the neurons it chemically connects are central to the erosion of memory and intellect. A massive effort has been under way to devise methods of restoring acetylcholine to its neuron synapses. Aricept is the current drug of choice; it acts by inhibiting the enzyme that breaks down acetylcholine, thereby lubricating failing circuits.

The positive effects of Aricept are variable. Using the maximum dosage, Morris notices that the quality of his writing is enhanced. He is less impressed by what the drug can do for his short-term memory. Far from a panacea, Aricept escalates some aspects of cognition in some patients. And it often controls the behavioral difficulties that arise. It is at once a stopgap and a promise that improved therapies are at hand for a disease whose name evokes the bleakest expectations.

PEOPLE WITH A TERMINAL illness must think in terms of caretakers—devoted family members, friends, the hospice to help them through the pending incapacity and death. Although they identify with those who share their specific illness in support groups, it seems improbable that two terminal people would allow themselves to fall in love, to live together; there are too many problems and uncertainties that would be compounded.

For Morris and Laura, whose affair seems unique in the annals of AD, the benefits of their shared lives outweigh the impediments. In their quest for meaning, they must continually reconfigure themselves, align their shrinking intellect, their muted emotions, their heightened spirituality, with their biological state.

"I think it is Laura who is the avatar of quiet heroism," Morris writes in his e-mail from Montana. "Living with incurable, progressive dementia, the horrors of the fact and the illness combined, is like living in the aftermath of a nuclear war, surrounded on all sides by devastation and waiting for the radiation sickness to make you finally wither and drop. Most of the survivors are lying on the ground, moaning piteously. A few are rushing about hysterically. One is gathering firewood, calling a support group meeting, organizing tasks, placing a comforting hand on a child's shoulder. That one is Laura."

Morris's insights and bon mots now await me each morning via e-mail. If I have a thought or question that intrigues him, it can spawn many notes. "It may seem complicated for us to live and love together here, but right now it's all pretty ordinary. We'll probably be scaling Everest, so to speak, in a few years."

Without reliable memories, Morris and Laura live in the present. They *experience* life and limit the rationalizing of it, seeking harmony rather than mastery in all that they do. I try to picture their newfound domain—the snowbound farm they describe at the base of cold, majestic mountains. I imagine them engaged in their own computer workspaces, communicating, as they do, with a variety of early-AD people. They are teachers and inspirations to the many who read their writings. Laura puts out a journal—lively discourses on what can go right or wrong in the span of an early-AD day. Morris's shared thoughts tend to be more academic, infused with the published writings of others.

"Laura and I both feel the blunting of our emotions, and we tend not to imagine or remember unless highly motivated," Morris writes. "Since Laura no longer drives, I do more of the errands and she does more of the dishes than she otherwise might. Watching a 'mindless' sitcom is too mentally taxing to be fun, and we seldom converse and linger over dinner because food takes the blood from our brains. When we want to share thoughts in a leisurely way, we lie on our bed in the dark, usually holding hands. This is not so we can be especially tender, it is so we can compensate for our vulnerability to fatigue and circuit overload."

SOME NOTES FROM LAURA'S ONLINE JOURNAL:

Sat. Jan 29: A word about certain vegetables . . . I was cooking and couldn't remember what that green one in the pot was called. Looks similar to a white one . . . broccoli. We agree that is the most likely spelling. We decided that it wouldn't be serious to confuse broccoli and cauliflower but to confuse peas and carrots would be a bad sign. Goodnight all . . .

Sun. Jan. 30: To clarify last night's "broccoli incident" . . . It wasn't that I couldn't remember its name. It was that I couldn't remember what it WAS. Its "broccoliness" was missing. Another cool morning, around 0°F. Goodnight all . . .

Fri. Feb. 11: I'm happy to say the broccoli has retained its identity. Had some tonight in a teriyaki dish. The unsettling thing about cognitive lapses, as I think of them, is that they are totally unexpected. And no way to prevent them happening as far as I know. They tend to stick in my memory because of the emotional content. It's not that I have a heartfelt connection to broccoli. It's the loss of its identity that's heartfelt. Goodnight all . . .

Wed. Feb. 16: Thanks Morris for getting the digital camera for me . . . a great gift. Once I master the learning curve I'll take a picture of the house addition and post it. The downside of the day was another fall . . . nothing serious but very unexpected. Sometimes it feels like my feet are no longer connected to my brain—just momentarily but at times that's long enough for something to happen. The snow and ice are melting. That should help. Goodnight all . . .

Sat. Feb. 19: Today makes a full year of taking Aricept. I'm happy that it is still helping me nearly as much as it did back six months or so ago. I see a lot of progress made during the year. This website created and one for Caregiver's Army . . . and my relationship with Morris as well as the continuing success of the e-mail lists I manage . . . I have lots to be thankful for and I feel blessed in many ways . . . Goodnight all . . .

Sun. Feb. 27: Morris is writing up a storm on various topics. Oh, this morning I woke up around five with lots of ideas on various topics and areas of my life. Wrote them down as fast as I could . . . I haven't had this 'flow' of ideas for quite some time. It's encouraging to have it happen again . . . Goodnight all . . .

—

THERE ARE THOSE with AD, like Laura and Morris, who retain an unusual degree of function. They seem spirited rather than apathetic, mindful to the extent they are allowed, and oriented toward solving the variety of problems their illness puts in their way. Terminal illnesses of the brain can have an advantage over those of other body parts in that the patient, so long as he or she is aware, is compelled to focus on the mind. With AD, the stark realities of neuron losses, the synapse depletions of neurotransmitter substances, a clogging of the neuroapparatus by tangles and plaques, close in around patients like an aperture. They feel this decreasing light as silent, painless curtailments of emotion and thought. This mental immersion, this raptness—neurons attuned to the plight of other neurons—can be the very source of their insights and inner strengths, even as an eerie darkness falls.

"THURSDAY, MARCH 2. I blew an MMSE (Mini Mental State Examination) question for the first time. I knew this had to happen someday," Morris writes. "Every year, the average patient loses three MMSE points from a perfect thirty. Typically ten years from diagnosis until death. Time is beginning to run out. The best I could answer for the day of the week was Tuesday or Wednesday.

"Afterwards, I searched through the cloudy past for a day/date I remembered and came up with Monday, December 20, the date my grandson Alistair was born. I remembered that it was Monday because the Caesarian had been scheduled for right after the weekend and I had driven a thousand miles to be there. I recall my elation on that very special day.

"Knowing that December 20 was a Monday, I can figure December 31 was a Friday. So were January 28 and February 25. March 2 was a Thursday. Got it! I solved the problem, but does it really matter?"

"IT'S NOW FOUR MONTHS since Laura and I have been a couple and it's time to deepen our commitment," Morris writes. "Neither of us have had a great deal of success in past relationships. Finally, we have with each other a chance to do it right."

Their life together is new, a closely shared experiment of love and exotic afflictions, a process of growth, a planning for death. Are they really much different from the rest of us? Are we "normals" not "a little demented" for the way we have managed to frame the human condition?

"At bottom, we're just a couple like any other, dealing with the challenge of having an I–Thou rather than an I–It relationship," Morris writes. "What difference does a little dementia really make, when the greatest minds have struggled in vain to know themselves or others?"

I FLY TO Bozeman, Montana, in early June and drive east to Paradise Valley, near Livingston. It is dusk when I arrive at the twenty-acre farm nestled in the snowcapped Absaroka Mountains; it is hay land irrigated with water from the Yellowstone River. Giant willow trees—a tree house sits in the largest—tower above the two farmhouses. Roy the Border collie welcomes me with his red Frisbee, communicating clearly with yelps and body language how he wants me to spend my time. An Arab mare named Rose and a quarter-horse mare named Serena are being courted in the pasture behind the houses by a visiting Morgan stallion named Zeus. He has sixty days to get these mares pregnant. (Only fifteen days left and counting.) Assorted domestic shorthair cats seem to pop up from farmyard crevices; they have adapted to Roy's exuberant intensity.

Laura's brother Frank owns the farm and he spends his summers in the old log farmhouse with his wife, Karen. Laura lives in the newer house and has converted it into a duplex to accommodate Morris. I meet Laura at dinner that night. She is tall and stalwart-looking, her blond curls emphatic and her hazel eyes trichromatic. Initially, she weighs her words carefully. But as she senses my respect and admiration for the way she and Morris conduct their lives, she mellows out to a kindly effervescence. I soon learn that their earlier intimacy has waned; they've settled into an evolving friendship, a shared journey.

In the morning, I visit with Morris. His new space—a study with floor-to-ceiling books, a living room with a counter for a small refrigerator and microwave oven, and a bathroom—smells new. A hallway with a door at the end connects him to Laura. The living room and hallway walls are bare, but this spareness is offset by windows that look out on rich, grassy pastures and the mountains beyond.

"I try to avoid a visual overload by keeping the rooms simple," Morris tells me. "The mountains outside project a feeling of realness that is satisfying to be in touch with."

Our Ventura, California, visit together several months ago is a dim memory for him, but a memory nonetheless; for many with early AD, memories of such a recent past are lost entirely. He continues to experience a gradual decline of visual intensity, short-term memory, and heretofore strong long-term memories—a linear progression of illness.

Paradoxically, Morris has increased his competencies. His IQ has risen. He believes that he has instinctively developed a capacity for getting to core issues while letting go of the more peripheral ones that may bring on a mental overload. Three years ago he could accommodate some of these peripheral matters, but he can no longer do so; he filters out what is not essential. Perceptually, he runs a tight ship. The noncritical issues float obscurely in his mind, and if he chooses to deal with any one of them, he will shoot from the hip; his response may be inaccurate, but it is largely an inconsequential inaccuracy. "It is like having two different personalities," Morris says, "one meticulous that I can call upon when required to, the other laid-back, energy-saving."

He gives me an example: He carefully scrutinizes all the numbers on his checks but pays little heed to the inconsistencies of his signature. To me, Morris's triaging of all that his senses parade before him keeps him in the moment. He needs this allocation system to properly function. As if in recognition of his courageous inventiveness, Morris's essay, "Potential for Rehabilitation in Alzheimer's Disease," has been accepted for presentation at the Pivotal Research portion of the World Alzheimer Congress 2000 in Washington, D.C.

I leave Morris for the moment and enter the doorway to Laura's complex. Her space is crammed with memory-supportive pictures and mementos, life-affirming plants. Her thirteen-year-old blind cat, Chrissie, a veteran of years of RV travel, is always at her side. Unlike Morris, Laura does not seem concerned with visual overload but speaks instead of her increasing distance from the shore of remembered life.

"Before Aricept, my loss of memories and failure to build new ones made me dread turning off the lights at night," Laura says. "In darkness, I could feel the bed beneath me, the bedcovers, my head on the pillow. Beyond those feelings was a void. Tears would come as emptiness stretched out and closed in around me."

It is the absence of memories, the loss of context and connection that startles Laura. And it is her emotional reaction to it that enables her to remember each individual gap that plagues her. In addition to her loss of "broccoliness" and "geraniumness," she recalls an episode when she lost the sense of her son Mike. "I saw a figure approaching the house. It looked something like my son Mike, but not unequivocally so," Laura says. "His 'Mikeness' was absent. I stood by helplessly as he approached. I was too embarrassed to ask him who he was. After all, I'm not a blind woman. He finally said 'Hi, Mom,' and I knew it was my Mike, but I never forget how I temporarily lost my sense of 'Mikeness.' What terrifies me is the accumulation of these blank spots. Will they slowly build until they disconnect me from my life?"

I try to conceive of a mind so perforated with memory gaps that it severs a discerning woman from her life. The Aricept helps Laura hold on to her thoughts and backtrack more easily to previous ones. And yet she has mixed feelings about the upcoming celebration of her mother's eightieth birthday in Canada. Laura's three children and four grandchildren will be there, but she realizes that the entire event may dissipate in her mind like the wake from a speeding motorboat. At the very least, she will always have her notes to read, and pictures and videos to see some snatches of this special family day. And she prays that at birthday-party time, her mind will not lose any of her loved ones' essences. The anticipation of this six-day visit with her mother, of quality time spent with her, is partially sapped by what can befall her.

Morris now joins us to reflect on the complexities of their relationship. "We are synergistic insofar as we stimulate each other's conversations and give each other ideas."

Laura agrees. "We talk a lot about our AD, comparing our experiences," she says. "It helps us define and clarify the parameters of our illness."

But they both deal with the cumulation of their AD deficiencies. This makes living together harder and tempers the synergy. The future will bring more reliance on each other, but as their neediness increases so will their unreliability. They are resigned to the fact that they will eventually need caretakers, and realize that outside help can also be synergistic; it can simplify the mechanics of their existence while allowing room for their own creativity and collaboration.

Their illness enlarges them, and almost everything they say or do is thoughtful. In their work with AD others, they study those with more advanced disease and try to figure out how they would handle the various impairments. It encourages them to discover that they still have abundant cerebral reserves.

Laura remains a perky optimist who laughs easily. She manages two e-mail communities—CPWML for people coping with personal memory loss and Caregiver's Army. She also logs her daily journal online and is a chat-room host for PlanetRx, where she encourages AD patient socialization and mental stimulation.

Morris will deliver his paper at the World Alzheimer Congress next month. And he plans to collaborate with another writer on a book about his illness; what he and Laura live through each day will likely be the fodder for it.

Their lives together are unique and meritorious. With intelligence and wit, Morris and Laura seek to process their thoughts for as long as they can, despite diminishing circuit capacities. In the end, their minds will slow to a halt, their nerve cells no longer meaningfully connected. With the coming of newer drugs, that time may be years away. That is their fervent hope. And after spending an exhilarating weekend with both of them, it is mine, too.

IN ITALO CALVINO'S MEMO on "Lightness" from his book *Six Memos for the Next Millennium,* he observes how the lightness in life is permeated by the "Ineluctible Weight of Living. . . . Perhaps only the liveliness and mobility of the intelligence escape this sentence." It is precisely the mind's loss of liveliness and mobility, its infusion with insupportable heaviness, that define AD.

As we rush about, processing our thoughts like Mad Hatters, it behooves us to fix upon our fragile webs of neurons and know that there are diseases that would take our minds from us. For a time during the slow, inexorable course of AD losses, some patients can still concentrate, compensate, reformulate, and through a more leisurely, linear thought processing, live lives of remarkable merit.

Let us not forget.

CHAPTER SIX

The Burden of Sickle Cells

AT SURGICAL PATHOLOGY CONFERENCES, cells are the focus, not patients. Lecturers project photomicrographs of tissue slides onto large screens so that everyone can see the magnified images of cells; it is part of the learning process to sit in the dark and witness unusual cellular features, patterns, dramas that illustrate the lecture. A trained eye can readily identify the rogues and heroes, the plot twists. Although these are stories about the mechanisms of human disease, they seldom explore the cages of human illness.

In an effort to stay tuned to the larger implications of my surgical pathology practice, I attend clinical gatherings and listen to the findings of internists, surgeons, or pediatricians. At one such conference on sickle cell disease at the Hyatt Regency in Long Beach, California, I am surprised to see mothers of sickle cell children. Though I have never before seen patients' families at medical meetings, it seems to me a compassionate strategy. Aspiring to learn about the plight of sickle cell disease, I introduce myself to one of the mothers, Kim Holland, and her affected son, Comille Begnaud.

Comille suffers the pain of his disease about every second day, Kim tells me as we sit down to lunch and wait for our food. It is capricious pain and often intractable, usually arising in his chest, back, belly, or joints with varying intensity. On occasion, the pain can occur in his head, even his penis. Seemingly no part of him is spared.

"If the pain is mild, he gets that look on his face and his eyes drop real low," Kim says. "He stops playing Nintendo, or playing outside, and lays around the house. I ask him if he wants his pain medicine because I know how he's hurting, but he usually says no."

"Why don't you take your pain medicine, Comille?" I ask. He takes Motrin, or Tylenol with codeine if the pain becomes more severe.

Kim sits between us, so Comille leans in to her shoulder, talks to me across her chest. His rich chocolate eyes—the color of his skin—are enhanced by steel-rimmed glasses precariously perched on his tiny, upturned nose.

"I don't want to be a baby," he says.

"He's trying to learn to deal with the pain by himself, without medicine. So he lays down with a hot pad instead," Kim says. She has a pretty, angular face that contrasts with Comille's softer, broader features. "When the pain is severe and doesn't go away with the medicine, he comes crying to me. I know he can't take it anymore. He's grunting and he'll let out a yell." She is speaking faster now and louder as if it is her pain, too. But her cells have contributed only one defective gene to Comille's recessive disease; she has the sickle cell trait and remains pain-free. The other defective gene had to come from his long-absent father. "I panic because I know that he needs a morphine drip, that he has to go to the hospital. But he's calm as he cries. He tells me it's okay, he's just hurting. *He's* calming *me* down. I try to be strong for him and he's being strong for me."

"Women aren't strong. It's the men who are strong," Comille says. He is eight years old.

And pain is not the only leitmotif in this disease; disordered immunity allows seemingly harmless infections to fulminate. Comille has been hospitalized more than fifty times in his short life for pain or fever or both.

His outward appearance, like so many sicklers, is unremarkable; this makes it harder to garner sympathy and compounds his predicament. But the smile on his face is so ebullient that I am caught up in it, and I grapple for a way to help.

"Have you ever seen sickle cells?" I ask.

Kim and Comille look surprised.

"Under a microscope."

They shake their heads. "No."

"Come to my office sometime. I'll show them to you."

"Wow!" Comille says. "Can we go?"

Kim nods her approval. The thought of seeing the cause of Comille's illness has never occurred to them; nobody has ever suggested it.

Will the sight of his sickle cells brace him against his illness? I do not know. I can only show him these tiny aberrances in the best possible light.

OF ALL THE HUMAN CELLS I see, none is more pragmatic than the mature red blood cell. Pierce your finger with a sterile lancet and steer one oozing, sanguine drop onto a clean glass slide. Smear this precious grume thinly and evenly and, when it has dried, stain it with Wright-Giemsa. Place the slide on a microscope stage and behold an abundant uniformity of mature red blood cells. They are orange-colored disks, biconcave as Rolaids, as near and apart from one another as checkers on a board.

To purify their purpose, these cells discard their nuclei like unwanted tumors and stuff the newfound space with added hemoglobin. This pigment binds oxygen from the lungs and releases it to every vital body part. The binding and releasing of oxygen is the red cell's raison d'être; each cell travels some 700 miles in its 120-day life span, coursing the body's ubiquitous vascular network, discharging its vital element. Red cells travel easily; in larger arteries and veins where they mass, they are carried in the course of plasma like fry teeming in a stream; in the narrowest capillaries, they readily deform to half their diameter and squeeze through single file. Pliancy enables these acrobatics, keeps the blood flowing. Sometimes I close my eyes and meditate: Aware of my breathing—feeling my breath come in . . . feeling my breath go out—I picture the billions of red cells percolating my system, and try to tune in to their harmony.

So perfectly adapted is the red blood cell and so vital its function that the smallest flaw can be destructive. A single point mutation in the hemoglobin molecule is responsible for Comille's pain. The normal hemoglobin A in each red cell is virtually replaced by an abnormal hemoglobin S in sickle cell disease. When red cells filled with hemoglobin S release their oxygen, hemoglobin polymers build slowly and stretch the cells like kites into the rigid, crescent shape of a sickle.

Within the asylum of small vessels, younger sickle cells and vessel-lining endothelial cells can run amok. Aided by biologic modifiers, they can wax adhesive, embrace in torrid intravascular tangos, and narrow vessel lumens. In larger vessels, lining cells are less inclined to take part in the dance. They burgeon as wallflowers, thicken vessel walls, and also narrow lumens. When older rigid sickle cells are trapped in these dwindling vascular spaces, the blood can clot.

And blood clot deprives distal tissues of oxygen. It is the fierce pain of this deprival that can squeeze the spleen, heart, bones, muscles, liver, and lungs like a vise until sickle cell patients howl.

I have looked at sickle cells in countless blood smears over the years and never once have I thought about the pain; it starts to enforce its limitations within months after birth, to propel its host through a guarded childhood into a wary adolescence. And the longer the host prevails, the more organ damage from oxygen deprivation he sustains, until organ failure or infection or both cause a premature demise. The average male lives only forty-two years, the average female forty-eight; if the fierce distractions of illness are subtracted from their lives, their allotted time is much less still.

Reynolds Price, in his affecting memoir, *A Whole New Life,* describes a postoperative bout as "one long fling at the end of a tether of pain." When I think about Comille's "long fling," how pain can rise inside him, I see his body as a fire and his sickle cells as hematologic arsonists. I imagine Comille's early memories filled with pain, repeated and random intrusions that fire his cranial wires until they flare in the heat of it.

"COMILLE IS TEASED in class because he's so short," Kim tells me the following morning. We are standing in the hotel lobby drinking coffee, awaiting the start of the conference's second day. Comille's sickle cells have a reduced life span, and the chronic anemia that results can delay physical maturation, retard growth.

"Mickey Washington, the man that lives with me, makes Comille laugh at his shortness," Kim says. "He'll tell Comille to walk under the car and see if anything's leaking."

"I can ride free on the bus," Comille says. He is laughing as though he is part of the joke. "I can get into movies at half price."

There is something else he can do. Act. His shortness allows him to play a five-year-old, gives him an edge in the TV-commercial business. He has made a peanut butter commercial that airs next month on the Fox network. When the producer asked him if he wanted to work in commercials, he said, "Yep, I need a job."

But pain crises and infections do not allow Comille to live his life in an ordinary eight-year-old way; they obligate him to focus deeper. Some days, when this deep focus depresses him, he tells Mickey to "take good care of my mom." Such talk scares his mother. With Mickey, she tries to accede to Comille's great burden, help him cope with the flagrant unfairness of it. Most days Comille seems resolute; he can also be idealistic, infinitely practical. "I want to be a doctor when I grow up," he tells me. "I want to cure sickle cell disease."

MINE IS A RESEARCH-QUALITY German microscope made by Zeiss. It occupies the center of my desk and, truth be told, the center of my life. I fiddle gently with the coarse- and fine-adjustment knobs that sit like ears above its rectangular foot, run my fingers up the elegant curve of its arm to its flattened, mobile nose where finely ground lenses hang like walrus tusks. I hold the binocular body in my hands and search the eyes; in them, if I look deeply, I see malignancies that undermine their hosts, benignancies that whelm them with joy. My microscope and I are intimates; together, microcosms open to me with awesome clarity. I feel like a virtuoso exhorting sweet notes from a grand piano, or a stunt pilot at the helm of his aircraft, artfully furling contrails.

Today, to my right, sits my microscope's little sister, a Wesco monocular student microscope with ample resolution; it is Christmastime and I have purchased it for Comille. Before he arrives with Kim and Mickey, I place a slide of sickle cells on his microscope stage, set the pointer on a classic crescent.

"Apples and bananas," Comille shouts as he looks through the eyepiece of his new Wesco. He is looking through the high-power objective—a magnification of four hundred; his little hands deftly work the adjustment knobs, focus in on the shapes of his pain.

Mickey Washington, a gentle man, sits quietly in his wheelchair. A few years back, he unknowingly drove his car into a Hispanic gang's turf. The shooters never said a word; he was a black man and that was reason enough. A bullet pierced his spine at the fourth thoracic vertebra, made him a paraplegic. Kim sits by his side, holding two-year-old Kentrel, who is free of sickle cell disease.

"Suns and moons," Comille continues. He is excited as he sits at his new microscope, carefully taking the measure of the misshapen, miscreant red cells that so reduce the quality of his life.

"What do you say to Dr. Nadler, Comille?" Kim says.

"Thank you very much." Then he says, "What are the big cells?"

With the pointer I show him the different white cells: Lymphocytes have round purple nuclei and wear their skimpy cerulean cytoplasm like short skirts; they are responsible for immunity. Granulocytes have multilobated lilac nuclei and faint pink cytoplasm peppered with granules; they are the first lines of defense at the site of bacterial infections.

"It's the red cell bananas and crescent moons that cause your pain and infections," I say.

"Yeah. They look kinda weird. Wow!"

Perhaps he will visualize, in his own creative way, what his sickle cells can do to him. Can these flaws beget a positive purpose, a grace?

Now Kim sits at the Wesco; she, too, is excited to see the real thing. Then she tells me about her own paraplegia. While she and her four-year-old daughter (Comille's older sister) were walking to school in 1991, they were hit by a speeding car that had run a red light. Her daughter was killed. Kim was left a paraplegic and remained that way for a year and a half until regaining her ability to walk. Comille was two years old when he lost his sister.

I am struck by the violence that shapes their lives. For Mickey and Kim the destruction is external, random, abrupt; for Comille it is internal, ongoing, Mendelian. Yet there is a cohesiveness that shapes this family, a pride they seem to acquire in one another's company. They handle the gravest vitiations in their lives with quiet confidence.

ABOUT ONE IN TEN Black Americans has the sickle cell trait and one in four hundred has sickle cell disease. People of Hispanic, Mediterranean, Indian, and Middle Eastern origin are also afflicted. The only cure for this disease to date is a bone marrow transplant. In theory, this procedure is a godsend; the patient's marrow is destroyed with drugs and replaced with that of a normal, genetically matched sibling. The sad reality is that genetically matched siblings are not available to most sicklers. And when a match *is* available, there may be a lack of financial or psychosocial support; at times, parents and doctors themselves refuse this treatment option, fearing that it is still too experimental, that the patient is not sick enough to take the transplant risk. A marrow transplant can fail despite a genetic match. The patient can die.

For Comille, marrow transplant from a sibling is not an option; he no longer has a sister whose marrow might be compatible. Kentrel is adopted.

GROWING UP IN CANADA, I thought the difference between blacks and whites was simply a matter of color. If one color seemed unnatural or inefficient, it was my own pink-white that produced insufficient melanin to protect me from the sun's ultraviolet rays; I would simply turn red and peel.

In training, when I first glimpsed black skin under the microscope, I readily saw protective granules of melanin pigment tucked into the basal layer of the epidermis. I subsequently saw the magnification of my own white skin; melanin granules were barely visible, and the pigment-synthesizing melanocytes seemed languid.

The number of melanocytes in human skin is relatively constant, irrespective of race; it is the amount of melanin these cells synthesize that largely determines skin color. Black is just like white, but with more of nature's sunshade dispersed in it.

In my practice I see, in skin biopsies, the endless variety of normal pigment scatter; it is my job to see beyond this pigment, to focus in on the real conundrums in the biopsy—inflammation, benign and malignant proliferations. This is what surgical pathologists do. It is only in the larger world where every eye can spot melanin disparities that such differences are intuited as larger differences.

And what of Comille? Most of the sickle cell patients at his clinic are black, the doctors white. Kim is aware that some of the black mothers distrust white doctors, but she is grateful for the care Comille gets. And she sees bigger issues—a lack of awareness of sickle cell disease, insufficient testing for it by hemoglobin electrophoresis, inadequate genetic counseling, and paltry fund-raising.

I receive the following e-mail:

Dear Dr Nadler

this is Comille. I will go to beauty and the beast tomorrow at 11 in the morning then I will go with my mom . . . to take some tests that Dr Groncy ordered. they wont hurt its just a lot. And I like having all my doctors. I dont care if they are black or white they are my friends and they love me like I love them. well I have to get off now because I am supposed to do my homework and study from 3–4 then I can play on the computer until 8 and then it's time to wind down (that's what my mom calls it) and then go to bed at 9. My mom says I have to keep a regular routine. ok I have to go before she gets back or she takes 30 minutes off my computer time. bye bye Dr Nadler I love you.

Comille

IT IS A CURIOUS THING that an organ configured so largely of nerve cells—the brain—itself is without nerve endings for pain. Within the sanctum of its vessels, sickle cell clots can deprive nerve cells of oxygen, painlessly ravage them. These attacks (strokes) can come on silently, without warning, as vague as the haze of a morphine drip.

Is Comille at risk for such a cataclysm? Psychological tests show a discrepancy between his high IQ and his reading, visual memory, and visual processing skills. Such disparities in a sickle cell child whose nervous system is otherwise intact suggest a narrowing of his cerebral arteries, possible early brain damage.

In the black-and-white images of magnetic resonance and ultrasound, the narrowed curlicues of his cerebral arteries are confirmed. Although his brain appears as yet undamaged, the diminished blood flow (and oxygen) it is receiving puts Comille at risk for a stroke.

It is possible to reverse these vascular changes, to reopen his lumens and resume a full blood flow. The solution is not simple.

THE PALE BLUE CURTAINS and linoleum in the daycare room at Long Beach Memorial Medical Center teem with multicolored polka dots, the white paper walls with pastel-colored fish. At the far end, an array of toys—building blocks, trains, trucks, puzzles, crayons, stencils, soldiers, Barbie dolls—is scattered about. Giant video games hug the wall. TV screens face downward from elevated brackets. It is the olive-green vinyl reclining chairs, spread uniformly across the floor, each with its own IV pole, that give the room its medical purpose: to infuse chemical agents into children with cancer, to transfuse blood into children with thalassemia and sickle cell disease.

Comille is the first patient to arrive. His short hair is done up in tiny twisters, his smile guarded. He will require a unit of packed red blood cells every three to four weeks of his life to markedly reduce his chance of a stroke. He has had occasional blood transfusions before when his hemoglobin got too low, and remembers how good the infusion of normal red cells and the elevation of his hemoglobin made him feel.

He climbs onto the first reclining chair with Kim and Mickey by his side. The nurse has difficulty sticking an IV catheter into a vein in his left hand. He is calm, his face inscrutable, his hand gilded with blood. The nurse apologizes, seems flustered as she cleans his skin and connects him to a packet of sodium chloride solution that hangs from the IV pole; it will flush his vein, keep it open to receive the foreign blood. She secures the catheter in place and supports his hand with a splint.

When the bloody poke is over, Comille climbs off the recliner and wheels his IV pole over to the toys. I tell Kim and Mickey of the mounting evidence that repeated blood transfusions in sickle cell children with narrowed brain vessels may not only prevent a stroke, they may, in a small number of these kids, cause the narrowed brain vessels to appear reopened.

They nod. The clinicians have already informed them of these hopeful possibilities.

I am reluctant to tell them about the risks of repeated blood transfusions, the immunization to foreign red cell antigens, the iron overload in body tissues that can occur. As a surgical pathologist, it is not my place to discuss therapeutic complications. I try to draw the line between demonstrating the microscopic alterations of a disease—the apples and bananas of it—and clinical management. I hope the depictions themselves, the vivid red cell aberrances, will help mollify, for Comille, his disease's hard edge. Pain is more than a stimulus transmitted by nerves; it is a complex perception largely influenced by emotion. If the sources of his pain—sickle cells—can be perceived as positive images, perhaps he will feel more assured in his illness, less fearful, his pain more tolerable.

I have been so engrossed—a unit of cross-matched blood now hangs on Comille's IV pole and trickles through the catheter into his hand vein—that I have failed to notice the other children. Now I see six of them cheerfully interacting and moving about, each tethered to an IV pole, each receiving chemotherapy or a unit of blood. They are playing with the toys, eating their lunch—macaroni and cheese, ham sandwiches, potato chips, chocolate pudding—laughing, seemingly oblivious of their ongoing therapies. The day-care center has the feel of a giant pit stop on the road of life; refueling and maintenance checks are a necessary routine, a fragile ingenuity.

The hours go by and Comille remains in a Super Nintendo trance: Ever so slowly, normal blood seeps into his blood vessels, diluting the sickle cells, sufficiently elevating his hemoglobin to turn off the marrow's flow of new sickle cells. Sitting beside him as though tethered to his pole, I am awed by this stay of his illness, swept up in his new and congruous flow.

It is still uncertain how sickle cells bring about clotting, how vessels narrow, how chronic transfusions seem to reopen the narrowings. These murky complexities add mystery to the misery of sickle cell disease. And yet there is hope: In a decade's time, perhaps more, Comille will donate the stem cells of his marrow to genetic engineers; these cells will be grown in tissue culture, spliced with normal genes, and returned to him. Cured at last, his marrow incapable of making new sickle cells, he will be given a hug and sent on his way.

In the meantime, Comille endures his disease. He dutifully gets his blood transfusions and lives his life as best he can. And we keep in touch.

I receive another e-mail:

Dear Dr. Nadler,

I got a lot of blood at the hospital and I didn't leave until late at night. The next day I was dizzy at school and got rushed to the hospital because the doctors thought I could have a stroke but I didn't. they kept me over night to watch me and I was scared because everything got really dark but I was strong for my mom. I can't wait for the Torch Run and I can't wait to run with you it will be fun. How is your wife and dogs please tell them I said hello and my mom says hello and Mickey too. well I will go now to Dr. Groncy's office for my checkup.
I love you Dr. Nadler. Your friend forever.

Comille

The old *Queen Mary* has been docked at Pier J in the port of Long Beach for twenty-eight years, functioning as a hotel and museum. Beneath its imposing bow is the newly completed Queen Mary Events Park, the sight of this year's Torch Run. I go with my wife, as I regularly do, to honor the valor of children with cancer and blood disorders, to help raise funds to defray medical expenses. It is a grassy park packed with children, parents, siblings, friends, dogs, all participants wearing the kelly-green Torch Run T-shirts. Each sick child, in turn, with torch in hand and parents and sponsors in tow, runs, walks, or rolls in a wheelchair a quarter lap around the track. It is a victory lap for all, and each winner is celebrated with loud kudos and the vigorous applause of spectators thronging the track. These are spiritual moments as kids get to momentarily forget their illnesses in a positive environment.

It is Comille's turn to carry the torch. He explodes from the starting line, the torch held high in his right hand. His little legs pump like pistons as the crowd cheers. He is soaring now, pushing his limits, bravely confronting misguided genes. I jog happily in his wake, his pride playing over me, and know that I will sponsor him until he is cured.

CHAPTER SEVEN

An Old Soldier

THE HOUSES IN SAM PATTERSON'S CALIFORNIA neighborhood are much like the ones in my own, contemporary stucco and wood structures, painted in pastels, trimmed with white, and aligned in tight rows on small lots. His is powder blue.

As I walk up the inclined ramp to his open front door, Sam wheels himself out to greet me. He is tanned and burly, with gray, kinky hair and an ingratiating smile. His stock-still legs seem as redundant as those of a puppet. He raises his right hand and shakes mine vigorously. Then he ushers me through a beige, tiled foyer into a mahogany-paneled den, to the straightback chair beside him.

The house has a musty smell and the light is muted by uniform beige walls, mahogany woods, and coffee-colored shag carpets in every room. This visual hush displaces the outside world and ensconces me. The ambiance is in sharp contrast to the pastel-colored enamel walls, the harshly reflected light, and the din of veterans' hospital wards where Sam has spent much of his life.

Phyllis Page, a colleague of mine and Sam's physician, has encouraged this visit; she is convinced that some good will come from an acknowledgment of his story. Until my visits with Patti Fleming and Mehli Mehta, I had not made a house call in nearly thirty years of practice. I remain cautious. Lacking clinical experience—for years I have hidden myself behind eyepieces, objectives, and condensers, separated myself from others with the accumulated glass of tissue slides—I suspect that I will observe too much, sacrifice microscopic clarity for the greater clinical scope.

"I was a professional tap dancer before entering the navy in 1941," Sam says. "Served four years as a Navy Air Corps pilot and never got so much as a scratch on me." His agile hands play as he speaks, as though he is signing, and he modulates his voice like an avid storyteller. "So in 1946, after the war, I volunteered to check out fighter planes and bombers at Yanabaru Air Force base on Okinawa. In the evenings, my buddy and I would tap-dance in the CATS [civilian actress technicians shows] there, and on the surrounding islands. We *wowed* those crowds."

As I listen, quickly taken by his effervescence, I can almost hear the rapid-fire tap-dancing, feel the syncopated rhythms, see his eager face glisten in the humid Asian heat.

"I was making touch-and-go landings one day in a Privateer with two of my buddies," he says. "I set the plane down on the starboard landing gear to test the shocks. The damn flight tower never warns me that the nosewheel is flapping in the breeze. So we skid along the runway trying to rev up, to raise the nose again." He pauses, shakes his head. "No can do. At the end of the runway, the tail flips in a partial nose-over and I crash against the bulkhead, gashing my left leg." He runs a finger from his mid-thigh to his hip to show me.

"An ambulance comes to take me to the field hospital at Yontan," he says. I barely hear the quiver in his voice. "We're on the road a few minutes when the left front tire of the ambulance blows and it rolls off a cliff. I fly through the rear doors, ripping up the back of my head and my left shoulder."

He is momentarily silent, consumed by memories. Then, almost matter-of-factly, he tells me that he tore his spinal cord apart at the level of the seventh thoracic vertebra.

I can almost see that vertebra being forcefully dislodged from its vertical alignment, severing the enclosed spinal cord with the abrupt finality of a guillotine. He was twenty.

Confined to his wheelchair for more than fifty years, Sam doesn't seem worn by it. Though fate had forever changed him that day on Okinawa, I can't help wondering, as I listen, how he has handled it, whether or not he has healed his soul.

"I'm not sick, I'm paralyzed," he says. "And people don't understand that a paraplegic loses more than the use of his legs. His bladder and bowels get altered and he loses the feeling in his skin." He carefully points out the skin of his lower back, buttocks, abdomen, and legs that have lost sensation.

"And your sex life is never the same. My three wives can tell you something about that." He grins sheepishly.

With motor paralysis and sensory deficit in the genital area, Sam's erections are reflexive and unrelated to erotic stimuli; they are involuntary and give him no sexual gratification. Despite this, he talks about a satisfying sex life.

"When I was a younger paraplegic, I got extreme pleasure from lip and neck contact . . . and nipples," he says. "It was as if God had granted me a new kind of orgasm." He cocks his thick gray brows as though confiding these details evokes the carnal textures and odors, the exhilaration of it.

Though ejaculation is primarily a genital event, orgasm is primarily a cerebral one. Is it possible that orgasms are adaptable, that different neural circuits can be used to achieve similar results? Could Sam's sex drive have forged alternate circuits to achieve his purpose? Although the adult brain seems to alter its neuron circuits in response to its environment, there are, as yet, no specific answers to these questions. I begin to feel the uncomfortable magnitude of Sam's clinical complexities.

Then he grasps the handles of his chair and, with arms still powerful, like a gymnast on rings, gracefully lifts his body, holds its dead weight high. This relieves the pressure on his buttocks, allows sufficient blood to flow, prevents pressure ulcers. It seems as though a minute passes before he slowly lowers his body to its former sedentary position.

We sit quietly for a moment. Sam's gaze is cordial. Perhaps he is musing on the fact that he is lucky to be alive. He belongs to that first generation of spinal cord injury (SCI) vets who survived their injuries. Prior to World War II and the development of sulfanilamides and antibiotics, the majority of people with injuries like his died from complications of kidney infections and skin pressure ulcers.

Sam and the SCI vets who survived with him lost their purpose when the Fascists surrendered and they languished in veterans' hospitals, consigned to oblivion for years. They were war heroes whose isolation from society was surely greater than most of the inmates in our penal institutions. And the management of their injuries was primitive. They had to learn from one another and from a few dedicated doctors and nurses. Despite the magnitude of their experiences, they were seldom nurtured; hospitals were not—and are not still—designed to integrate technologies with matters of the heart and spirit.

Sam backs his wheelchair up against the center of a low-lying couch behind us. Several pillows are stacked there, and as he raises the foot of his wheelchair, leans back, and locks the wheels in this position, his body rests on the pillows. Creatively, he lightens the load on his buttocks.

On this couch, to his left, is a radio, telephone, clock, pad and pen, current *TV Guide,* answering machine, Sony Walkman with earphones, calendar, skin cream, perfume, and a picture of Bob Dole. To his right a ten-pound weight and the TV remote control. The couch is Sam's base of operations; I suspect it is here that he achieves the freestanding coherences that feed his independence.

He lifts his ten-pound weight and begins to do biceps curls. It is essential to maintain his upper-body strength in order to smoothly execute his one-hundred-and-seventy-pound body transfers, he tells me. I try to conjure his mighty load, his arms and shoulders transferring his thickset body from place to place in the course of his day—bed to wheelchair to toilet to wheelchair to car to wheelchair to toilet to wheelchair.

"I left the hospital in May of 1952," Sam says. "I finally realized that dependency is a lack of freedom. Life was passing me by." As he talks passionately about the emergence of his new life, how he has consolidated himself, his wins and losses, I envision, microscopically, the severed edge of his spinal cord, stained with a silver impregnation. It is years after the injury, and once-vital sensory and motor fiber tracts and collections of nerve cell bodies are largely replaced by bands of scar tissue. A few residual nerve cells, their black, scraggly tendrils splayed like Einstein's hair, have managed to survive. Bound in scar, they appear like resourceful weeds poking through cement cracks. I also see small benign tumors (neuromas) bound in the scar. Dysfunctional severed nerve fibers, trying to reconnect, have formed hopelessly matted skeins.

Although I have seen microscopic sections of old severed spinal cords, I have never given any thought to the human embodiment of them. Sam's electrochemical nerve impulses were completely disrupted at the T7 level; they have never again been transmitted or received below that level. His lower trunk and limbs, his bowels, bladder, and genitals, are permanently incommunicado, shutting him off from the rest of his body like a demented mind.

Primo Levi writes in *The Drowned and the Saved* that "the absence of signals is itself a signal, but an ambiguous one, and ambiguity generates anxiety and suspicion." I wonder what anxieties or suspicions Sam's absent spinal signals have generated in him. I ask him.

"People pity me, ignore me, or stare at me, as if I'm different somehow." Sam is doing incline presses for his pecs. "They never really see me; they only see my wheelchair."

I think about the paraplegics I see in the course of my day, in the hospital corridors, on elevators, in restaurants. They are wheelchair people, slowly climbing and descending ramps, pulling and pushing their way into and out of vans, craning their necks to the sky to converse with the likes of me. True to Sam's words, the burdens of their electrochemical disconnections have never occurred to me.

"I went back to school and became a counselor, learned to work with SCI vets and, later, with handicapped kids. Ah, those kids." He vigorously slaps his knee with his hand, fixes me with an indulgent gaze. "They've given me back my life! I used to see them here, as many as fifteen in an afternoon. Sometimes I had to go out after them, bail them out of trouble."

For most of us, our bowels and bladder work discreetly unless malfunction announces their presence. For the paraplegic, they are conspicuous for the lack of communicating nerve impulses, and I ask Sam to tell me more about the hold they have on him.

"You won't know about my private functions without seeing my bathroom. Hell, you're a doc. It shouldn't bother *you.*" His eyes twinkle, as if reflecting his moxie.

I follow him to his bathroom, not sure what to expect. At first glance, it appears ordinary; it is small and tiled in white. Then I notice the toilet tucked into the far corner, the bowl set right against the wall so Sam can slide himself onto it with greater ease. The seat is padded with soft white vinyl to prevent his hip bones from compressing his buttocks. It is raised six inches above the bowl by bolts and wing nuts. A horizontal metal bar runs three feet above, fastened to the wall stud. Its free end is protectively covered by a punctured tennis ball. The whole contraption has the aura of a Rube Goldberg machine.

Sam positions himself beside the toilet, grabs the bar with his hands, and raises himself out of his wheelchair. Sliding smoothly onto the vinyl seat, he centers himself and, in order to keep from falling off—his abdominal muscles are paralyzed—places a plastic jar wrapped with padding between himself and the wall. His body transfer complete, he places a foot-long wooden leg spreader between his thighs and pulls a latex rubber glove onto his right hand. "I'll go through the motions, show you the way I defecate, if you'd like."

I nod. What else can I do?

"If this was for real, I'd take off my pants and dip my gloved finger into that jar of K-Y jelly." It sits on an accessible shelf in front of him. "Let's assume that I've done it. Now here's what I do," he says. "Watch me."

With his left hand back on the bar to secure himself, he whips his gloved finger around to his backside and sticks it through the six-inch space between the seat and bowl.

"See how I get my finger to my anus? Every night at eleven o'clock, I gently rim it with the K-Y jelly and slip in two glycerin suppositories."

There is not a hint of embarrassment as he shows me his moves. Though he is a long way from tap dancing, he seems to be giving an inspired performance. He is a cynosure in that bathroom, so effective with the viable half of himself that I almost feel physically redundant.

"Eleven to twelve minutes later, I gently massage my anus," he says. "Then I defecate just like anybody else. If it doesn't come out easily, I dig it out with my finger. No big deal."

I try to imagine Sam's required intimacy with his movements, how it must be a continual reminder of his profound accommodation.

Sam has to be careful: People with severe spinal cord injuries lose voluntary muscle control of their bowels; if they survive through the years, they have hemorrhoids from periodic episodes of constipation. If Sam massages his external sphincter too vigorously, he bleeds. He has to regulate his meals, his fiber intake, his vertical posturing to have normal bowel habits, to avoid embarrassing accidents.

"When I'm done, I always wash my privates," he says. "I keep the leg-spreader in place, soap up, and pour cups of warm water over me so it runs into the toilet." He pauses and looks at me for a reaction. "Want me to go on?"

I nod once more.

"I dry myself and powder Wilbur. Then I cover him with tincture of benzoin and attach a gizmo [external urinary drainage device]. I like to put a new gizmo on every day. Less chance of a leak that way." He shakes his head and his face clouds with sadness. "So many SCI vets I see at the hospital stink of urine. They get tired of paying so much attention to their needs."

With complete loss of voiding reflexes and bladder sensation, Sam's control of urination is deranged. He lowers his pants and I see the gizmo; a condom with its attached latex rubber tubing is fixed to his penis by the tincture of benzoin and a liquid skin adhesive on the shaft. A single layer of elastoplast, wrapped onto the condom just below the penile head, secures it. Nothing is inserted into the penis.

"If the elastoplast is too tight and I have an erection, Wilbur can strangulate," he says. "If it's too loose, the gizmo comes off and urine soaks my pants."

Getting Wilbur and the gizmo to function in sync has required a surgical cut through the bladder's external sphincter so that urine dribbles like a slow leak. The alternative—thrice-daily insertions of a bladder catheter—enhances the chance of urine reflux and infections and would further reduce Sam's independence.

Running along the outside of his atrophied leg is the latex rubber tubing that connects the condom to a urinal bag at the calf. "I keep three of these ducks

[bags] and I'm constantly disinfecting them," he says. "Once you get in the habit of properly hooking up the gizmo, it's all very routine. Believe me."

Then his chest sags as if it has sprung a leak. "It's old age that's sapping me," he says.

What I see in the bathroom is a pattern of physical limitations, each augmenting the other, each adding to Sam's burden. I had no idea how much skill and patience a paraplegic has to acquire, the constant energy he has to expend to execute his bodily functions. It is remarkable how a determined mind can pull its body along.

A WEEK GOES BY. Once again Sam awaits me at his front door. His eyes blaze up at me as he eagerly grasps my hand. As we sit in his den, I realize that it was the canny mechanics of his paraplegia and his personal magnetism that distracted me from the signs of his age. Now I note his age-spotted, wrinkled skin. Lacking in turgor, it looks loose as molt. I picture his skin biopsy, the dermis laden with degraded collagen, a homogeneous, gray-blue sprawl. And his gnarled hands. A biopsy of such arthritic configurations shows eroded joint cartilage, burnished and thickened underlying bone, even joint mice (dislodged pieces of bone and cartilage) that mar its spaces.

I ask him how aging affects his paraplegia.

Sam hesitates. "My old age has got me into a struggle at the Veterans' Hospital that makes piloting in wars and living with paralysis seem unimportant."

He speaks deliberately, trying to control his emotion.

"Hospitals anger me, the constant cost-cutting and downsizing," Sam continues. "That SCI unit is my second home. They look after me when I'm not right, and the older I get, the more care I need. Now the government is replacing our experienced registered nurses with inexperienced vocational nurses. They float from ward to ward and don't even know how to do a trunk turn or a body transfer." He squints his eyes and purses his lips, seeming to shape his streaming thoughts. "The aides we used to have were experienced and wonderful. They've retired, and the floating aides we get now usually ask *us* what to do. And many of our best SCI doctors, with years of experience, are leaving." He sighs, as if the isolation his body defines for him never seems to wane. "Even the equipment is archaic and needs repair," he mutters.

He is wheeling himself in front of me, in long

arcs, back and forth, like a sheepdog working his flock. I'm not sure if this movement is a nervous pacing or an exercise to sustain his strength or tone. I do know that the lightness of my first visit is gone.

Old SCI vets like Sam have overused their upper bodies; the ceaseless transfers have taken their toll. He tells me that his shoulders and elbows and wrists are arthritic, that the rotator cuff muscles and tendons in his shoulders are inflamed and the median nerves at his wrists are trapped in their carpal tunnels. Then he swallows, and his larynx extrudes its Adam's apple like a punch. "This is not the time to be dismantling the unit," he says. "It will dismantle *us*."

I certainly see the same sad havoc he speaks of in other hospitals. But I realize that old SCI vets represent special circumstances. For fifty years Sam's resourcefulness and self-reliance appear to have kept up his self-esteem; I saw it when I first visited. But age has crept up on him and all the body transfers that he has made without a thought in earlier days are becoming struggles. A proper SCI unit helps keep Sam's upper body fit, enables him to go on transferring his own deadweight for as long as he is able. I suspect this strength is closely associated with his will to live.

I RETURN TO MY microscope and gaze at a slide of old severed spinal cord. For as long as I can remember, I have witnessed the autonomous growth of malignancy, the variegated spread of inflammation, the organized cellularity of benign proliferations. But the neurological defacement I see before me results from forces external to the microscopic process. I can almost feel the violence of it.

I try to picture Sam's severed nerve cells gradually sprouting, regenerating into an experimental graft of peripheral nerves, perhaps a purified Schwann cell implant. The sprouts advance through minefields of growth inhibitors and scar, determined to bridge the gap. As dogged as he, they seek to reconnect with long-lost organs, to enable him to walk again. I am so close to these sprouts—the length of a microscope tunnel separates us—that I am caught up in Sam's dream, cheering his neural progress.

Then I realize I am in the wrong dream, that dreams of walking are for younger SCI patients like Christopher Reeve. Older ones, like Sam, have lived so much of their lives with their dysfunctions that accommodation is a greater reality than walking. *They* probably dream of a lasting ability to transfer; unhinged from gravity, they float from place to place like butterflies and die with dignity before they are grounded.

CHAPTER EIGHT

Dying Matters

IT IS CURIOUS HOW DEATH SIDLES UP TO us, dispatching some before they can settle in to life, catching up with others only after the longest lifetimes. A similar disparity occurs with our cells: Those that compose our skin or line our digestive tract are quickly sloughed into the environment or gut; others live longer, and nerve and muscle cells live as long as we do. Since our cells seem programmed to die at different rates, how can we know our body's true age? Gerontologist Leonard Hayflick suggests that we are as old as our oldest cells, that a birthday really celebrates the age of our nerve and muscle cells.

As a surgical pathologist, I am a daily presence at cell death: If it is death from injury or disease, inflammatory cells often gather around to mourn; if the death is DNA-programmed and has no visible cause, seldom does a single other cell appear to take note. No matter. These multiple daily minideaths I see build inside me until I am hardened to them.

Despite my acquired emotional detachment from dead and dying cells, I must be wary of them to be competent. I focus on what may be lurking behind these tiny corpses, looking beyond them for causal organisms, deficient blood flow, ever ominous cancer cells. I am aware that dead and dying cells are often merely clues left behind by an omnipotent cause: My rooting out a disease's origin leads to treatments and cures. By emphasizing cause in this way, I can relegate cell death to the background, in effect, deny it.

Almost daily, I stare down the facets of disease, cellular tales of woe that have a hand in human dying. Despite this exposure, the final phase of cell life is as mysterious to me as human dying. Cells will darken their nuclei, shrivel their cytoplasm, become opaque as they die; but deciphering their opacity—the reality of their dying—is beyond me. This awesome complexity falls largely within the realm of cell biologists these days. Their video-enhanced microscopes allow them to see the live interplay of cytoplasmic organelles, to fathom the meshing of all these tiny engines that drive us.

Although my work is hospital-based and eighty percent of people in this country die in hospitals, I seldom visit the terminally ill for purposes of a needle biopsy or aspiration. On occasion, in anticipation of a surgical biopsy, I go up to the patient floors or the special-care units to review a chart. Seeing patients all tubed and taped there, I am often struck by the flagrant loss of their privacy, how the impact of all that is impersonal is magnified. It is the sight of someone's life ebbing in such a place, with fear or anger or humiliation on display, that stays with me; somehow these patients seem less than the sum of their cells.

The word "aging" is a euphemism for dying. And though our decay teaches us in very visual and perceptual ways how our bodies begin to close down, we choose to deny its lessons. We spend billions of dollars each year cosmeticizing ourselves: we dye our gray hair, adorn our balding scalps with toupees, fill in the clefts of our wrinkles, lift our faces, and suck out our fat. Until the decline of our bodies directly threatens our lives, we forever fret over them, chemically or surgically alter the visible testaments of age, and never contemplate dying. Preoccupation with veneer distances us from death, making it more extraordinary than it really is; it shrouds death in myth, in the fear of excruciation and humiliation.

Our cells, however, seem to accept their decay in a natural manner: Lipofuscin, the granular pigment that accumulates in aging heart and liver cells, is proudly flaunted around time-worn nuclei like golden necklaces; and nuclear faces, age-altered, seem not to effect the slightest pretense.

I SIT IN THE steam room at the local YMCA with an old friend. Alone and naked, we talk about death.

"Each of our lives is a story in search of an author. When I listen to dying people, I am less concerned about their character than what kind of a character they are, and how they have populated their lives with other characters," Brad Deford says. He is a middle-aged pastor who gave up his congregation to become a chaplain—a spiritual and grief counselor—

at the local hospice; it is attached to the hospital where I work. "The meaning of people's lives then becomes evident in a narrative construct."

I inhale the steam until it coddles my bronchial byways. "How do you elicit the stories of people's lives?" I ask.

He laughs. "Most of us spend our lives in solution, where we appear very similar to one another, and we fear that our lives don't mean much to anybody." He is a large, muscular man who so often laughs at what he says that it is easy to miss the sweep of his conceptions. I listen with care. "I try to be the precipitant, to precipitate out the stories that are already there and allow for meaning in life that is clear and embraceable."

I think it must take courage to enter a dying person's story, to press buttons here and there, to fix on a denouement and allow for a revelation. Brad's role is not a passive one, but it is not active either; he does not *edit* a dying person's story; in the course of conversation, he facilitates it, allows it to come forth of its own volition.

Steam hovers thick and white until we can barely see each other. The room is edenic, or is it hadean?

Our lungs, now laden with steam, sigh in exhalation. The heat and humidity balance our humors, purge us.

It was Michel de Montaigne who said that "he who would teach men to die would teach them to live." If hospice counselors like Brad bring meaning to people's dying, make them see the value in their lives, perhaps we need this kind of gentle affirmation *throughout* our lives. I tell him so.

"Our stories—the ones that give meaning to our lives—can best be told when we accept our death," he says. "Unfortunately, for most of us, this doesn't occur until shortly before we die." Only his words are audible now; steam-weighted, they ride the air until they fall like drops on my exposed surfaces. "An intuition, without obsessing about it, that death could come at any time is an acceptance of it; such an intuition propels us to edit and reedit our stories until we get to the deepest truths," he says. "Death then becomes the long-term goal that sets our life in motion. Intensifies it. Enhances it."

"IT'S FEAR THAT ENCUMBERS dying, isn't it?" I say. I have joined Brad for an early-morning coffee at the local Starbucks. The caffeine seems to rouse those around us until we can hardly hear each other above the din.

"Often so. When it's crunch time, we're quite likely to be greatly surprised by what we fear most, even ashamed by it." His easy smile gives the seriousness of his words a gentle edge. "When dying people tell me their fears, it often takes everything in me not to provide some easy reassurance," he says. "It's very hard to hold fear with someone because everything in me cries out to say 'Let's find a practical way to address this.' But if we rationalize too much about our fears as we die, we are not accepting death."

How Brad cares for the dying, how he helps people come to creative, if not peaceful, ends by working the narrative kinks out, seems inspired; life stories get crafted from snippets of bliss and woe, hope and yearning, anxiety and despair.

"Death is not a particularly rational happening, and much of what we have not known about ourselves gets exposed in the process." He scratches his thick brown beard with its shards of gray, and his eyes widen as if to reflect all they have seen. "Sometimes it is best to listen and hear a patient; sometimes it helps to go deeper," he says. "There are moments when we come to wordless understandings, often with tears in our eyes, that dying is not something that can be made better or made to go away."

I can't help but wonder how many tears he has shed in the throes of all these deathbed cogitations.

"I have to have good inner boundaries to continually take into myself and feel the sorrows and fears of others, and not become overwhelmed," Brad says. "I know I'll be surprised by my own fears when my time comes, and I believe that listening to others helps me prepare my own possible future."

He seems to practice his own dying with every patient he sees. I wonder, by filling his life in this way, with an inchoate sense of his death, how much he comes to savor the reward of each granted moment.

WE DRIVE TO THE home of a man who died the day before to visit with his grieving family. En route, Brad tells me how he empties himself of expectations and whatever it is he thinks he knows before each house call. We park in front of a small wood and stucco structure painted yellow and white. A large maple tree towers above it. It is late fall, and brown and orange deciduous leaves cover the lawn like a vibrant shawl. As we stand at the door, two sizable, well-disposed fellows, I feel a little like an apprentice door-to-door Bible salesman.

"Meet Dr. Nadler," Brad says as we enter the house. "He's shadowing me these days to witness the spirit of hospice."

The widow—I'll call her Mrs. Jones—is there with a woman friend. We are ushered into the living room where light-colored walls and fabrics counter the dark wood floors and furniture. The women, both lean and tall and middle-aged, are in surprisingly good spirits. Mrs. Jones quickly credits the hospice nurse and volunteers with her husband's peaceful demise. She tells us how relieved he was when they entrusted the family with his care, instructed them in the control of his pain with drugs, and enabled him to die in his home with easy access to family and friends.

"After my husband died, the body remained here for five hours," she says. "It gave us a chance to let go of him slowly." I try to picture the body, the bone and flesh that Mr. Jones left behind. This embodiment of her husband allowed Mrs. Jones to slowly and gently back away from his sphere.

Brad engages her in conversation. She talks about her future plans as though she is already acclimating to her loss. She tells him that her father died at this time of year twenty-five years ago, how her husband's death has enabled her to make peace with her father's.

The purposeful serenity, the thoughtfulness that appears to have surrounded Mr. Jones's death, is in stark contrast to the medicalized dying that occurs in hospitals. Since dying is not a medical process, hospitals mostly devalue it, strip it of its potential. Despite their technical bravura, medical centers have mostly failed to weave the process of death into their fabric.

"This was a good death," Brad tells me in the car. "It fits with our sense of what we might hope for ourselves."

"And Mrs. Jones's apparent lack of grief is a part of this good death?"

"Ah. But she does grieve," Brad says. I am Dr. Watson to his Sherlock Holmes.

"Now twenty-five years older and wiser, she is more effective with the dying of her husband than she could ever have been with her father. So her present grief gives her the chance to rework and regrieve her father's death and come to a new peace with it."

This is the first of several cryptic or symbolically expressed communications I will miss as we visit the dying and their families in their homes.

DURING THE MEDIEVAL CRUSADES, a hospice was a resting place for weary pilgrims. Today it is the conceptual hub of a reactionary movement, a way of comforting and caring for the dying and newly bereaved that humanizes the process. It often helps patients conjure up the import of their lives by winnowing time-faded merits; and it teaches them to readjust their hope rather than lose it in the face of mounting losses.

"I never visit a dying patient's bedside to chatter idly. I try to find meaning in what they tell me, to help with deeper issues." We are sitting in Brad's tiny hospice office. The walls are covered with poems, pertinent articles and cartoons, bon mots, thank-you cards, diplomas. A picture of Mahler, his favorite composer, is on his desk, and above him is a Lichtenstein print of a beautiful woman speaking in a cartoon bubble: "I know how you feel, Brad."

His sounds like a practiced art to me, to find meaning in someone's life under the duress of dying, an intense companionship that helps to ease the bottled fear, to quell the somber rumblings.

Today we drive to the east side of Palos Verdes. A physicist I'll call Homer lives there with his wife. A three-pack-a-day smoker for more than thirty years, he has widespread lung cancer. Their daughter is visiting from Seattle for two weeks, and this gives Homer a window in which to die, to allow the daughter to help her frail mother with the death and all of its multifarious arrangements.

They are happy to see Brad, but there is something wrong with the scene: We sit in the den, to Homer's left; he is tucked into his recliner and faces a large rear window that looks out onto San Pedro Harbor; a giant-screen TV with its outdoor satellite dish and 257 channels is stationed beside him. We gaze at his distorted profile, his belly bloated with fluid, his flexed legs elephantine. Although his position is fetal, he stares out at the ocean with the solemn bead of a ship's captain.

"I see you have a new clock," Brad says. A large digital clock keeping military time and the date sits on a shelf to his right.

"Ah." Homer fixes Brad with lucid, bilious-green eyes. "It lets me know if it's day or night. Sometimes I doze off and get confused about that."

We sit and quietly allow him his observations.

"The line of zero longitude at Greenwich is the basis for calculating military time, you know," he says.

Brad's laugh is infectious yet gentle. "You sound as if you're going on a voyage."

"Not until I can handle all this." He chuckles.

"All this" refers to his illness. Homer has yet to accept its lethal nature. He is fighting for a cure in the face of insurmountable odds. A cannula attached to an oxygen cylinder is inserted into his nostrils and he leans forward, methodically smacking his buttocks to prevent bedsores. Then he leans back in the recliner and suddenly falls asleep.

"I've had to up the dose of morphine this week," his wife says as we leave the room. She is stooped and withered, as if her husband's dying is more than she can bear, yet she smiles inscrutably. As we retreat through the hallway and living room to the foyer to make our exit, I notice how all the paraphernalia of his illness—oxygen cylinders, cannula coils, oxygenators, boxes of medication—seem to clash with the low-slung wooden beams and country-style charm of the house.

"They don't yet accept his dying and have no wish to talk about it," Brad says when we leave. "Homer refuses a hospital bed that's far more appropriate to his care. As long as he's in that recliner, he's not a 'cancer patient' and doesn't have to deal with all that that implies."

The digital clock, the multichannel TV, the harbor view keep him tuned in to life. He is not yet ready to accept his oncoming death. Brad has no intention of drawing him into his dying. This man will have to find his own way. And if he comes to accept what must be (many die and never do), Brad will escort him, in amity and love, to the river and see him off to the other side.

The digital clock, its conveyance of military time, may have other implications here. Scientists are largely future-oriented and have greatly expanded time perspectives; they often seek to predict future occurrences by understanding past events. The clock may simply be reflecting the time-orientation maxim of this man's profession. But it ticks amid so many other denial clues that its scientific significance pales.

In nearly all cigarette smokers, cells that are harbingers of cancer arise in the lining of the bronchial tree that wends its way like roots into the porous loam of the lungs. Behold the angry-looking cells revised by perpetual puffs of smoke, their nuclei as deeply blue and broody as a stormy sky, their cytoplasm a fiery pink and recoiled as if in repudiation of all the abuse. In needle aspirations of lung cancers like the one that will soon take the life of this physicist, I often see intimate clusters of tumor cells, craggy nuclei molded together, uselessly merged. Tumor cells break away into the flow of blood and lymph, nestle in liver, adrenal glands, brain, bone, or anything else they can. Once ensconced in distant sites, they cannot be entirely extricated; they multiply to a critical mass and overwhelm the body unto death.

Bronchial lining cells will take only so much abuse before they turn on you.

THE FOLLOWING WEEK we head north through the beach cities to visit a biology teacher and her husband. Their house is cluttered with Christmas decorations (it is February). Wreaths, children's toys, decorative lights, and candles are everywhere. The husband is reluctant to remove these vestiges of his wife's final holiday.

The woman I'll call Barbara is marred by the spread of her breast cancer, and her wanton wastedness, her woe, tells all. Her hematoxylin-blue eyes hold a confident detachment as she awaits death with impatience. She wants to die, she calmly tells us. To get the thing over with. Living with dying has allowed her to make her final peace and it has no further meaning for her; all that she needs to move on to her hereafter is now within her.

Her husband brings her a peeled and sliced pear; with tremulous hands she slowly and disinterestedly consumes it.

"You don't look ready to die, Barbara," Brad says.

His words surprise me; she seems nirvanic, taking in her final breaths. Later he will tell me that eating and wearing lipstick are expressions of a persisting life force that is clashing with her will to die.

She focuses firmly on Brad when she hears his words. Slowly she tells him how she awoke at seven o'clock last evening thinking she had slept through the night. Finally realizing that it was only dinnertime, she was greatly disappointed that she had not put another meaningless day behind her, that she would have to face that day.

Brad sits by the head of Barbara's bed, as close to her as he can get. He observes and gently engages her as the husband talks to me about his book club.

"When we read *The Handmaid's Tale,* the Margaret Atwood book, I realized how the whole thing could work in reverse," the husband says. This is a curious observation: Atwood's "Handmaids" and "Hannas" are servants in this dystopian novel. Does this husband see his current role in such a demeaning way? Perhaps it is the dying itself and not servitude that is dystopian, the reversal referring to the change from what was utopian.

I realize that the husband needs attention, too. He is weary of his wife's dying, the intensity of it, the abrupt necessity of his loss. I let him gently vent while my attention is held by Brad's earnest gestures. Soon Barbara tells him she is in pain today, that she will be better able to talk of her feelings when next we visit. As we exit, the husband uses an eyedropper to drip morphine onto her tongue.

"You made things easier for me today by engaging the husband," Brad says to me in the car. "I was able to stay with Barbara, acknowledge the meaninglessness of her days. I could feel her comfort."

It is a paradox that cancer arising in such a readily accessible organ as the breast continues to exact such a heavy toll. In half of these patients, malignant cells have dispersed beyond their organ of origin by the time the tumor is discovered. It is often an insidious wandering, a spread to the remotest areas of the body. This woman's bones and flesh are now as seeded as pomegranates with cancer cells.

"Barbara's plight, her wish to be done with dying, is not uncommon," Brad says. "Too many of us have no idea how to make our last days and hours meaningful."

Having thrust myself into Barbara's life as an observer, I find it curious how her impasse becomes my own. Life stockpiles our griefs and losses, and as we die, still greater losses are added to the inventory. The loss of our sense of self—even our basic bodily functions that we mastered at the age of two—must go very deep. I cannot conceive of a way to make this kind of experience meaningful.

"We are independent and active for most of our lives," I offer. We are sitting in late-afternoon traffic. Brad shifts in and out of lanes, but we don't advance; we are in our own impasse. "I think that aging has to teach us to adapt, to trust and surrender to our inevitable decline," he says.

We talk about *Tuesdays with Morrie,* the bestselling book by Mitch Albom. In it, Albom rekindles his relationship with his dying college professor, Morrie Schwartz. "Morrie was able to find dying of undying interest and Barbara could not," Brad says. "The key was Morrie's willingness to be dependent. To enjoy this dependency. To revel in it. To return to a childlike state of being completely taken care of. Unconditional love. Unconditional attention." Albom concludes that as Morrie was dying, he "was giving as an adult and taking as a child."

When we visit the following week, Barbara lies in her bed embowed like a fetus, her hands contracted, her body twitching, her eyes tightly closed. The cancer is having its way with her, and she is no longer in communication with her surroundings. The time for her departure seems at hand. Her husband lovingly massages her feet as Brad and I sit quietly by. I cannot know the strength or complexity of the bond they are severing, the degree of their intimacy. I do know that the husband has been a constant at her bedside, that her children and grandchildren have been frequently there, that hospice nurses and volunteers have meshed with the family, according this woman the right to have control over her own death, in her own home, with hardly a falter.

Two days after our final visit, Barbara dies in her sleep. Despite the loss of meaning in her life in those final days, her dying is compelling to me; there is a peculiar beauty in the peace of it, in the family and hospice nurturing that quietly palliated her bodily discomforts. Is a beautiful death oxymoronic?

"Not at all," Brad says. "The love this woman cultivated in life was ardently returned to her in her dying. And there is beauty in that."

It's as though the bonds you form in the course of life raise the stakes of your denouement, move those around you in ways you can hardly conceive.

I'M SITTING IN THE Memory Chapel of the mortuary where Homer's body lies. To the end, he had never accepted his dying. It is evident from the large, respectful gathering of mostly gray-haired scientists and their wives that Homer's life had been far-reaching.

Brad is officiating. His tone is upbeat. He is clearly celebrating Homer's life. He asks those gathered to rise and tell their Homer stories, and a number of the gray-haired colleagues do. Resurrected from without, the sum of his endearments, Homer now reemerges larger than he was in life. Stories of him that reverberate in the neurons of people's memories keep Homer alive and testify to the value of his life. So will the stories live on in his children and grandchildren. We can never underestimate the sway—good or bad—our lives can have over others. Although some of us believe deeply in an afterlife, what *is* certain is the afterlife of our stories, how they reside in the cells of others, capable of enduring for as long as humans do.

THE HOSPICE PROTRUDES like a tiny appendage from the giant hospital much like meager Death pouts from robust Life. But the hospice movement is catching on, and people are dying outside of hospitals these days in ever growing numbers. One out of every seven deaths is currently bolstered by hospice. As baby boomers age, death will continue to leave hospitals to life and pervade, whenever possible, our communities. Hospice will then emerge from the hospital's shadow, bringing death out among us where we can see and feel and smell it before making it our own. All the impersonal movie and television deaths will have to compete with true, living versions of it. These living deaths will bring an ambrosial urgency to our lives, engross us with our finity. They will play havoc with our stagnation.

Carole Hoffman, a nurse I've known for many years, coordinates the sixty-one volunteers who are the heart of the hospice. They fill the caretaking gaps left by family and friends, and can be genial, talismanic presences to those who come alone to their dying. "They're young and old, professional and nonprofessional, working and retired. They all want to give something back," Carole says. "Lawyers, accountants, psychologists wear two hats here, their hospice volunteer hats and their pro bono professional ones."

I wonder how much wisdom these volunteers acquire and dispense about life, about death, by treating dying patients as human beings rather than people who are on the verge of death.

"I get to know volunteers intimately when I train them. This is how I sense who'll be right for each patient." A smile opens across Carole's face, anticipating her observation. "When I leave this job, I'm going into matchmaking," she says. "I've discovered skills I never knew I had."

Like Brad, she laughs easily in the course of conversation. Blood rushes to her head and her eyes fill with the kindest light. When I ask her to tell me how caregivers interact with the dying, she does so with stories.

"We had an engineer whose wife was dying. They had been married for fifty years," Carole says. "We were concerned because he wasn't expressing his feelings and wasn't letting any of us share his grief. But he made his wife a footboard and on it he wrote the date, day of the week, weather, and anything else he thought might be of interest to her. So whenever she woke, if he wasn't there, she could still be in touch with the world." This brief story moves me, the poignant creativity that can rise from a man's devotion. His tact was exquisite.

Carole shifts seamlessly to a new story. "One of our young woman volunteers is an attorney. I sent her to visit a dying old man who lived alone with his cat and wanted to draw up a will. He chose to leave all his money to his only friend and requested that this man be the executor. On his way to meet with the attorney, the friend keeled over dead. So the old man asked the attorney to be his executor and told her what he wanted done. He died himself a few weeks later."

These stories are filled with urgent sensibilities, earnest connections between moments, and Carole seems to have a deep satisfaction in the telling of them.

"As the executor, the attorney decided that a funeral was in order and telephoned me to bring some of the volunteers," she says. "First we all went to the old man's house to settle some of his affairs. There we met the little old lady next door who knew him, and we asked her to join us at the funeral."

"How many people were at the funeral?"

"Myself, three volunteers, the attorney, and the little old lady," Carole says. "During the service, four little old men enter the chapel and sit at the back. When the priest finishes and asks if anyone would like to speak, these men come to the front and, one by one, tell their stories."

Carole is laughing now as this funereal experience gathers narrative force, and I'm thinking that it is cynical to assign too much simplicity to anyone's life.

"They tell how they met the deceased at a card club in Gardena," Carole continues, "how he came to the club because he was lonely and liked to be around people. He didn't play cards, so he sat in a chair by the door and talked to everyone as they entered and left. In time, these five men, who previously hadn't known each other, became good friends, and the remaining four reflected on the difference the deceased had made in their lives. One man told how he never talked to a soul at the club, he had just played cards until the old man began greeting him at the door. Another told how the old man would always smile at him. 'What did he have to smile about? He was all alone.'

"By the time they finished, I came away feeling I knew more about this man than most I've lived with and buried." Carole pauses dramatically and I am struck by the extraordinary, how it can rise unexpectedly, even exhilaratingly, from the ordinary. "And it's all because of the love this attorney had for the old man and how she insisted that he have a funeral."

We are all lost in nature's great flow, but if we shape our own stories, if others help us configure them, our lives seem fraught with meaning, taken to a higher ground.

"A doctor contacts us about a middle-aged man who is dying with lung cancer," Carole says. "Our social worker and volunteer find him alone in his apartment. He's hungry and there's no food. He's so weak he has to crawl to the bathroom. When they call me I bring a commode and I run to McDonald's to get him a hamburger and a Coke. . . ."

WITH BRAD, I VISIT a Japanese-American woman who is dying of pancreatic cancer. Deep-seated, this tumor grew silently, its cells sprawling to adjacent, even distant, structures before symptoms appeared. It is fatal. Her prognosis notwithstanding, she has exceeded by many months her doctor's expectation.

As we enter their scrupulously tidy condominium we are greeted warmly by the patient and her sister. At a glance, I am not sure which of them is ill. They are both dressed in black slacks and printed blouses and bow graciously. Then I see that the hands of the woman I'll call Kiyoko are swollen, that her belly is too protuberant for her otherwise tiny frame, and she walks with the plodding shuffle of someone who is weakened. She is extremely polite, even jaunty, and as urgent as the taste of wasabi.

"It's almost time, Reverend Brad," she says with matter-of-fact friendliness as she ushers us into her living room. She is untwisted by disease.

"Yes," Brad says.

"When you speak at my memorial service, will you please keep it short?" she asks.

Brad's laugh is thunderous, and we all laugh, too.

"And not too flowery."

"You see how my reputation precedes me?" Brad says to me.

I am amazed at Kiyoko's candor, how clear-eyed she is about her impending death. She tells of her concern with the details of the memorial service and the cremation that will follow. She must choose what will be served at the tea, but she doesn't know how many people will be there. She frets about whether or not her relatives will be able to pick out some of her vertebrae and long bones after the cremation. As Brad will later observe, she has exquisitely separated that which endures from that which does not.

In her words and manners, Kiyoko masterfully incorporates death into life. She appears to have long ago accepted the inevitable and seeks to make her final phase of life as meaningful as she can to family and friends. She is completely committed to the moment.

She has concerns because she lacks a ritual framework with which to plan her death. In Japan, she remembers, a pagoda on wheels was used to transport the body therein to the crematorium. The vertebrae and long bones were picked out after the cremation, preserved for the family, and the mourners had tea at the crematorium. Here, in California, there is no pagoda on wheels and few funeral homes even simulate the Japanese funeral as she remembers it; the crematoriums pulverize all the bones and return only ashes; and the tea must take place at a restaurant. So Brad listens to all of Kiyoko's requests, the minute details she and her sister have spent hours working out, and reassures her that she can entrust all the particulars to others, down to the specifics of the tea menu. And he will honor her request to celebrate and remember her life in every way she desires.

Kiyoko, so alive in her dying, is not sidetracked by it; although she has little interest in the machinations of her dying cells, she is raptly engaged by her life that remains.

THE HEART CAN BE scarred by life, much as the person who harbors it. Ruddy myocytes, their uniform contractions propelling the blood through arboreal vessels with a wondrously steadfast beat, can be replaced, in time, by the rigid fibroblasts of scar. This gritty tissue repairs the heart rents of ischemia like a patch, keeping the surrounding myocytes tightly bound; but it cannot participate in the heart's forceful thrusts; the sheets of fibroblasts are too stiff. If scarring is excessive, so is the heart's inflexibility, and the loyal pump begins to fail.

A couple I'll call Lattie and Niles is the last housecall of our day. Lattie's heart, scarred beyond recognition, is failing her. At age eighty, it is barely able to keep her alive. Her inefficient contractions have caused a backup of blood, congesting her organs and ballooning out the chambers of her heart. Even the smallest increase in physical activity or emotional strife can overwhelm such a heart, cause it to fail. So Lattie lives a guarded existence, a temporizing that delays her heart failure (and death) and intensifies her fragile reality.

Like most people we visit, she is buoyed by Brad's presence. She is excited and animated as he sits down beside her, and I hope her heart is equal to the task. The coming to meaning at the end of life is a spiritual process, Brad often says, not a religious one. But Lattie is a religious Christian.

"Why am I so afraid to die now?" she asks.

"How much do you believe that God loves you?" Brad answers her query with one of his own. Such a question has little merit for someone like Kiyoko, but it goes to the heart of the matter with Lattie, who is a religious Christian.

She does not answer.

Last year, her husband Niles's Parkinson's disease was controlled by Sinemet, so she was not afraid of dying and leaving him physically bereft. But she lingered on, her spirit keeping her at the helm, a heroic coxswain to her oarsman-like myocytes, a racing shell of a heart that never said die.

Of late, Niles's rigidity is causing him to fall repeatedly. He is unable to get up, to care for himself without Lattie's help. In their married years together, they have become two nuclei in a single cell. Although Lattie's heart is rapidly tiring of the race, she is afraid to lose that race and leave Niles in his own deteriorating state.

They have come to hospice care as a binucleated cell. If Niles were to die before Lattie, she could easily accept her own death and die in peace. But she has less bravado now as her rapid deterioration makes it clear that she will precede her husband. Niles is the one-word answer to the poignant question she raises about her death fears. She struggles not to leave him in the morass of his disease, and so the myocytes of her heart keep up their desperate vigor.

HUMAN LIFE MUST TRANSCEND the lives of its cells, but the two are so inextricably bound that I cannot think about one and not the other. Cellular death is not always essential to cellular life. Harken back to the earliest bacteria; they reproduced asexually by binary fission, replicated their own DNA, and divided into two perfectly coequal clones of themselves. In this way they bypassed death. Bacteria today still reproduce in this manner and, in the absence of accidental death, are immortal.

With the evolution of sex and multicellular organisms, only germ cells—sperm and ova—came to retain the potential for immortality. Gerontologist Tom Kirkwood has postulated that all somatic (body) cells are evolutionarily dispensable. Although our somatic cells come together as the incarnate, thinking humans we are, Kirkwood suggests that our bodies serve primarily as vehicles for launching our germ cells into future generations. Hence the evolvement, after reproductive maturity, of senescence and death by natural selection. All somatic cells that escape death from disease or accident eventually age and "naturally" die.

Is transmission of our germ cell DNA from one generation to the next our lives' basic purpose? Will our somatic humanity—the sum of all our magnificent scientific and artistic endeavors—become more than busy footnotes in the grand and universal scheme of things?

Will we ever be as immortal as our germ cells?

ON HOME VISITS with Brad, I see different scenarios played out, note the quiet heroics of everyday people. At sixty-something I think about my own dying and grapple with the emotional arc of it. Witnessing others meet life's end in meaningful ways has made my own seem plausible and has me thinking about all that my cellular body may cease to be: I visualize my hundred trillion cells gradually dying and critical organs beginning to fail; myocytes of heart muscle ceasing their marathon of synchronized contractions; flowing cells of blood tumbling to a flagrant halt; mazy tiers of charged brain neurons pulsing with final thoughts. How awesome is this cellular ride, so steeped in mysterious efficiency. But it is human dying, so urgent and inevitable, that is graven into me.

Epilogue

I meet Comille and his mother at the hospital. It has been three years since I have written their story, and I am delighted to see that he is eight inches taller despite the presence of his sickle cells. Almost eleven years old, he has come to have his mother's lean and graceful look and seems poised on the edge of manhood. We sit and talk about their life with his continuing pain, their retrospective knowledge in dealing with it, their growth.

Kim's interpretations of the tilts of Comille's blood flow allow her to gauge the level of his fragility and either gear up for the more serious back pain or nearly relax when the pain in his hand or foot needs little more than sympathy and gentle massage.

Three years ago, morning pain resulted in Comille's loss of a school day. Now, unless it is severe, he brings his meds to school and covertly takes them. He is the only sickle cell child in his class and has no intention of getting teased. And he knows the importance of pushing fluids: Dehydration can precipitate a sickle cell crisis; ample hydration can abate his pain.

The approach to his pain management has also changed. He takes his medications when the initial pangs of his sickling strike. His doctors now believe that a rapid medication response blights the pain before it can blossom into a crisis.

Three years ago, Comille was taking supplemental classes in reading and math because he had missed so many days of school. Now he is in regular classes, in fifth grade at Mary Butler School for Performing Arts in Long Beach, with a GPA of 4.0. When he is at home in crisis, a teacher spends the day with him. If he is not up to reading or writing, she speaks his lessons to him.

"I didn't have a good relationship with my parents, and I swore that if I ever had a kid I'd be their friend before I'd try to be their parent," Kim says. "I've kept my word on that. Comille is my best buddy. He's beginning to go through the girl-phase thing these days, so he comes to me.

I want him to talk to me about anything."

THE WOMAN I CALLED Hanna Baylan lived with her breast cancer for many years and despite its terrible gravity—perhaps because of it—acquired a hard-won wisdom. Cancer cells finally chased her from her body, freed her of it, until all that remained was her soul and intellect. And she rose with this lightness above the weight of illness to touch all those who knew her.

MORE THAN TWO YEARS have passed since I first visited with the old soldier I called Sam Patterson, and I return to visit with him once again. Time has not had its way with him. He is still tanned and burly despite his seventy-five years, fifty-five of them spent in a wheelchair. After the small talk, he shows me his arm muscles, how they still bulge with an urgent liveliness, sculpted as a bodybuilder's, supple after decades of critical toil. Then he grasps the handles of his wheelchair and easily lifts his body, still holds its deadweight high to ease the pressure on his buttocks. It is a seasoned brawn that he shows me; each acutely delineated shoulder and arm muscle contracts with well-practiced efficiency. He takes unmistakable pride in his masterful, lengthy survival.

In the beginning, he excelled within the rigid confines of World War II. Then he found a way to trump the constraints of his paraplegia. And now, in his old age, he deals with the most harrowing restrictions of all: The medical care provided by his SCI unit at the Veterans' Hospital is so woefully lacking that some of the vets have instituted legal proceedings; in the course of these actions they have been informed that their medical records are missing.

There is a sharp flash of anger in Sam's eyes as he tells me these stories. For the first time in his life he is helpless. The enemy—inadequate medical care—is too nebulous, too much a product of bureaucratic ineptness for him to handle.

I assure him that I will always be there for him, that I will see to his care outside of Veterans' Hospital as it becomes necessary. And I realize that I am preparing to put into practice what I have been trained to do. To personally help people.

ABOUT THE AUTHOR

SPENCER NADLER was born and raised in Montreal. He attended medical school at Queen's University in Kingston, Ontario, before training in surgery at McGill University and pathology at Albert Einstein College of Medicine. He has practiced surgical pathology for more than twenty-five years in Southern California. His essays have appeared in *Harper's*, *The American Scholar, The Massachusetts Review*, *Cross Currents, The Missouri Review*, and *Reader's Digest. The Language of Cells* is his first book.

ABOUT THE TYPE

This book was set in New Caledonia, a typeface designed in 1939 by William Addison Dwiggins for the Merganthaler Linotype Company. Its name is the ancient Roman term for Scotland, because the face was intended to have a Scots-Roman flavor. Caledonia is considered to be a well-proportioned, businesslike face with little contrast between its thick and thin lines.